Funding
Clean Water

Books from
The Lincoln Institute of Land Policy

The Lincoln Institute of Land Policy is a school that offers intensive courses of instruction in the field of land economics and property taxation. The Institute provides a stimulating learning environment for students, policymakers, and administrators with challenging opportunities for research and publication. The goal of the Institute is to improve theory and practice in those fundamental areas of land policy that have significant impact on the lives and livelihood of all people.

Constitutions, Taxation, and Land Policy   Michael M. Bernard

Constitutions, Taxation, and Land Policy—Volume II
   Michael M. Bernard

Federal Tax Aspects of Open-Space Preservation   Kingsbury Browne

Taxation of Nonrenewable Resources   Albert M. Church

Conflicts over Resource Ownership   Albert M. Church

Taxation of Mineral Resources   Robert F. Conrad and R. Bryce Hool

World Congress on Land Policy, 1980   Edited by Matthew Cullen and
   Sharon Woolery

Land Readjustment   William A. Doebele

Measuring Profitability and Capital Costs   Edited by Daniel M. Holland

Incentive Zoning   Jerold S. Kayden

Building for Women   Edited by Suzanne Keller

Urban Land Policy for the 1980s   Edited by George Lefcoe

Fiscal Federalism and the Taxation of Natural Resources   Edited by
   Charles E. McLure, Jr., and Peter Mieszkowski

State Land-Use Planning and Regulation   Thomas G. Pelham

The Role of the State in Property Taxation   Edited by H. Clyde Reeves

Funding Clean Water   Edited by H. Clyde Reeves

Land-Office Business   Gary Sands

The Art of Valuation   Edited by Arlo Woolery

Funding
Clean Water

Edited by
H. Clyde Reeves
Lincoln Institute of Land Policy

with the assistance of
Scott Ellsworth

LexingtonBooks
D.C. Heath and Company
Lexington, Massachusetts
Toronto

Library of Congress Cataloging in Publication Data
Main entry under title:

Funding clean water.

    Papers presented at a conference hosted by the Tax Policy Roundtable of
the Lincoln Institute of Land Policy in Cambridge, Mass., Nov. 19–20, 1982.
    1. Water quality management—United States—Finance. 2. Water quality
management—United States—Costs. I. Reeves, H. Clyde (Herman Clyde),
1912-   . II. Ellsworth, Scott. III. Lincoln Institute of Land Policy.
Tax Policy Roundtable.
TS223.F86 1984   363.6'1   83–22202
ISBN 0–669–07409–8 (alk. paper)

Copyright © 1984 by D.C. Heath and Company

Published simultaneously in Canada

Printed in the United States of America

International Standard Book Number: 0–669–07409–8

Library of Congress Catalog Card Number: 83–22202

Contents

Preface and Acknowledgments   vii

Introduction   *Scott Ellsworth*   1

Chapter 1   **Clean Water: The Financing Gap**
George E. Peterson   13

Chapter 2   **Funding Clean Water: A Private-Sector Perspective**   *John W.L. White*   25

Chapter 3   **Government Subsidies for Clean Water: Who Wins and Who Loses**   *Stephen P. Coelen* and *William J. Carroll*   37

Chapter 4   **The Federal Role in Funding Clean Water**
J. Leonard Ledbetter   49

Chapter 5   **State and Local Roles in Funding Clean Water**
Robbi J. Savage   63

Chapter 6   **Funding Clean Water: The Los Angeles County Experience**   *Charles W. Carry* and *Robert P. Miele*   75

Chapter 7   **Funding Clean Water in Illinois**
Richard J. Carlson and *Roger A. Kanerva*   87

Chapter 8   **Funding Clean Water in Kentucky**
Jacqueline A. Swigart and *T. James Fries*   109

Chapter 9   **Funding Clean Water in Maryland**
Mavis Mann Reeves   123

Chapter 10   **Funding Clean Water in Michigan**
Doris Van Dam   139

Chapter 11   **Funding Clean Water in New York State**
Richard Torkelson   151

Chapter 12   **Funding Clean Water in South Carolina**
                   Clair P. Guess, Jr.   161

Chapter 13   **Funding Clean Water in Utah**
                   James O. Mason   169

Chapter 14   **Funding Clean Water in Washington**
                   Joan K. Thomas   187

Chapter 15   **Funding Clean Water in Wisconsin**
                   Paul N. Guthrie, Jr.   205

                   Tax Policy Roundtable Members   221

                   About the Contributors   223

                   About the Editors   227

Preface

The United States has enjoyed an abundance of clean water available at a low cost. But according to some writers and specialists, the nation's water future is very much in doubt—and, indeed, may soon develop into a crisis. Chemical pollutants and overconsumption aside, the core of the problem is the status of the nation's water infrastructure, that is, our clean water delivery systems and wastewater facilities. Many of these systems are inadequate or in a state of decay and will soon need extensive rehabilitation or replacement.

An awareness of the gravity of this problem, which carries a price tag of $2 trillion to $4 trillion, has come about at a time when the nation seems particularly ill-equipped to cope with it. How can the needed capital be generated in an era of fiscal austerity? Will it be possible in the future to adequately fund the water systems without seriously changing existing standards?

These and similar questions were addressed by a gathering of over thirty scholars, water specialists, and public administrators called together by the Tax Policy Roundtable of the Lincoln Institute of Land Policy in Cambridge, Massachusetts, on November 19-20, 1982. The Tax Policy Roundtable chose to explore the clean water problem because how questions about it are answered may have a lasting impact on tax policy and land use decisions.

The papers presented by participants of the conference are collected in this book. The introduction summarizes the discussion of the major topics and issues, which are explored more fully in the chapters that follow.

Chapters 1 through 5 provide a general perspective, whereas chapters 6 through 15 are in the nature of case studies and relate to the situations in particular areas. These case studies are arranged alphabetically by state.

Providing and delivering clean water presents different problems in most every region and community, and no list of case studies could raise all existing issues. Our inquiry has been constrained by what could be reasonably covered in a day and a half and has primarily focused on infrastructure, of which pipes are a big part. We deliberately avoided getting involved in environmental and water rights controversies, because we believe these important issues belong to different forums. The limited selection of case studies was substantially dictated by the pragmaticism of identifying knowledgeable people who would present the problem well. We think this goal was accomplished and that our sampling raises most of the issues. We view this effort as a start that others will expand with improved methodology.

In undertaking this inquiry in late 1982, after the almost simultaneous recognitions that the clean water infrastructure was in bad condition and that money to improve it was short, we knew that few clear answers would emerge. The dust must settle for vision to clear. It is never too early, however, to assess problems and ponder solutions. We hope this book will be helpful as the American people continue to assess and ponder their clean water problem.

The Tax Policy Roundtable is grateful to those who have contributed papers for this conference and also to R. Lisle Baker, Stephen Gold, Will Knedlik, E. Blaine Liner, and Reece Milkes, who participated as invited guests. A short bibliography of each contributor is included at the back of the book.

I wish particularly to recognize with appreciation the recording and editorial assistance of Scott Ellsworth and the typing and clerical assistance of Glenda Miller.

Introduction

Scott Ellsworth

On November 19, 1982, in Cambridge, Massachusetts, the Tax Policy Round-table of the Lincoln Institute of Land Policy hosted a conference on funding clean water. Members of the roundtable, the contributors to this book, and several invited guests were present.

Most of the papers read at the conference and subsequently published in this book were made available to the conference participants prior to the meeting. At the conference the authors gave only summaries of the first five papers, which allowed enough time for participants to interrupt the summaries with questions or comments. The summaries were followed by clarifying questions and answers. Discussion then became loosely structured but in keeping with an outline suggested by the chairman, which called for points to be illustrated by reference to the ten succeeding case study papers. The questions raised were far ranging, and agreement seemed to exist that, given the constraint of time, they were honestly and vigorously confronted. Discussion was often lively.

From a tape recording and my notes and memory, I have endeavored to provide a summary that identifies the major topics discussed and, I hope, fairly depicts the emphasis that prevailed.

Philosophies, Folk Values, and Definitions

Undergirding any proposal for financing U.S. water infrastructure needs is a set of important assumptions that are so basic that they are apt to get lost in even the most reasoned debate. They concern how we define water and how we feel about it. "Our feelings about water," conference Chairman H. Clyde Reeves stated, "are helping to create the environment in which water is paid for." Conference participants agreed with this assessment, and they gave considerable attention to how definitions of water affect any agenda for its future financing. "While our debate is how to pay for it," one participant said, "we first have to look at what it is, and what is its right of use."

In their attempts to fathom what might be termed an "American philosophy of water," a number of participants remarked on the differences between water and other utilities. "What is it that sets water apart from gas and electricity?" one participant asked. Several participants responded that differences in the public's attitude about water can be understood only in historic context—particularly one that takes account of the English commons heritage.

According to James O. Mason, executive director of the Utah Department of Health, the key distinction between water and other utilities lies in the fact that: "Gas companies turn off gas, and water companies don't. You can't live without water in our society. Water has to be available to people." Although John W.L. White, chairman and chief executive officer of Consumers Water Company of Portland, Maine, pointed out that his company does indeed in some instances turn off water for nonpayment, and although Paul Guthrie, director of Intergovernmental Programs of the Wisconsin Department of Natural Resources, stated that gas and electric companies are prohibited from shutting off services in his state during the winter months, it was generally conceded that Mason's distinction was an important one.

For a number of participants, the various attitudes of Americans toward water have been combined into an overall system of folk values. "In our country," stated one participant, "clean air and water are inalienable rights of citizenship." Federal Judge James Harry Michael also recognized the existence of such a general public philosophy of water, and he urged that future financing of the nation's water infrastructure would have to grapple with it. "Even with this great range of problems that we have facing us," he said, "we still have this American folk value that water is free. The American people recognize that it costs to distribute water, but they feel that water itself is free. . . . We must first concentrate on qualifying that folk value to make them see that water costs."

The overall pricing of water also received considerable attention at the conference. Here the discussion focused primarily on whether water has been underpriced, with many participants assenting that it was indeed. Jacqueline A. Swigart, secretary of the Kentucky Natural Resources and Environmental Protection Cabinet, added that "the public will support additional funding if—or when—it is necessary."

Last, conference participants also addressed the future of our water philosophy and the types of definitional issues that need to be addressed as we improve an often aging and inadequate infrastructure. One of the most important considerations involved special provisions for the needy in our future restructuring of water finance—something that many participants strongly urged be adopted. A second involved the relationship between water availability and growth. "Whoever controls water controls the land," Joan Thomas of the Washington State Department of Ecology remarked, while Stephen P. Coelen, director of the Massachusetts Development Research Institute at the University of Massachusetts, added that "the availability of water encourages people to live in certain areas." Some participants, arguing that a restriction of growth was necessary, felt that such availability may have to be curtailed in the future.

Availability of Water Dollars

Compared to other utilities and related services, average yearly water costs per capita in the United States are low, as can be seen in figures presented by J. Leonard Ledbetter, director of the Georgia Environmental Protection Division.

Utility Service	Cost per Year
Electricity	$438
Gas	380
Telephone	306
Garbage	97
Cable TV	228
Water	45
Wastewater	90

Nevertheless, as a number of conference participants pointed out, proponents of water bill increases may find that additional dollars from this source are difficult to find.

Inflation and the nation's current economic malaise have, of course, taken their toll, and they may continue to play a significant role in efforts to rehabilitate, improve, and extend clean water resources, delivery systems and wastewater treatment facilities. Not only has inflation eroded the purchasing power of all water dollars, but it has, according to one participant, "shot up the estimated costs for a completion of the Clean Water Act so much that it is evident that the federal government will not do the whole bill."

The current shortage of available public dollars will also have an immense effect on water-related construction projects. "What we all have in common," stated George Peterson of the Urban Institute, "is that water will be weighed against competing capital priorities." Efforts to boost water funding, other participants added, will face substantial problems along every step of the way. Robbi J. Savage, executive director of the Association of State and Interstate Water Pollution Control Administration, noted that water dollars generated through the user pay system often go into general public fund accounts and are "unable to get back" into a water program fund.

In the face of such an overall difficult situation, several participants urged a closer look into who currently pays America's water bills, with special attention given to the load carried by industry. Doris Van Dam, superintendent of Grand Haven-Spring Lake Wastewater in Grand Haven, Michigan, mentioned the case of one town whose leather tannery contributed such a wastewater load that the municipal sewage facility was essentially "an industrial treatment plant, plus a little public dilution." Although Van Dam mentioned that most plants charge on a volume basis, the plant she had referred to also paid in proportion to its pollution load (BOD and SS), biochemical oxygen demand and suspended solids. At the end of this discussion, a number of participants wondered whether industry has been paying its fair share.

According to Guthrie, industry in Milwaukee is not at a disadvantage when it comes to water bills. Approximately 60 percent of the flow in Milwaukee, Guthrie explained, came from six industries, a situation that led the city to build certain procedures into its treatment processes because of the unique character of the flow. When a study was conducted into who was paying for the city's overall treatment, however, it was determined that the industry enjoyed an overall advantage over other groups in the community. "The most disadvantaged were small businesses," Guthrie said, "then residents. Industry had the best situation."

Yet, regardless of whether the industrial sector is generally benefiting in its share of water costs, some participants urged the conferees not to forget by whom all the costs are ultimately borne: "The costs," Clair P. Guess, Jr., of Columbia, South Carolina, said "ultimately all go back to the consumer."

Pollution Standards

As noted, any future scenario for financing America's water infrastructure cannot limit itself to a simple tallying of how water dollars are to be procured and spent but must consider certain aspects of the public's perception of water. One of the most important—and sensitive—points that it must address involves pollution standards.

Conference opinion on this general area was mixed, though it tilted generally in the direction of environmentalist concern. While some participants were indeed critical of what they perceived to be unduly high water pollution standards, most participants expressed counter opinions, and many gave graphic evidence of some of the results of water pollution in America. Mavis Reeves, professor of government and politics, University of Maryland, noted how the pollution of the Chesapeake Bay had done much to cripple Maryland's domestic fishing industry, and Richard Torkelson of the New York State Department of Environmental Conservation explained that Long Island's uppermost aquifer was showing signs of increased organic contamination, forcing the state government to close down numerous water wells on the island. Others noted the increase of toxic wastes that are finding their way into America's waterways, and Guess reminded the conference that "there are more and more toxic waste compounds that can't be unmanufactured." Most appeared to agree with one participant who stated that "there are more incentives for industries to pollute under the current [Reagan] administration."

The conference also witnessed some division of its ranks on the issue of who should be responsible for establishing water pollution standards. Some urged more local control, but the majority seemed to opt for national standards. Ocean dumping was cited by a couple of participants as an area where federal pollution guidelines were particularly needed.

Closely related to the issue of local versus national pollution standards was that of "pollution havens"—their existence, or lack of, and what should be done about them. "There is some concern that a few states won't come up to snuff," noted Savage, alluding to the concerns that many northeastern and midwestern industrial states have that Sunbelt states will steal their industries by the lowering of the southern states' pollution standards. While such concerns may seem well grounded, a few participants voiced their doubts whether such pollution havens currently exist. "I just don't see them that much" remarked Ledbetter.

It should be noted, however, that in this as in other areas, patterns among the states are not always predictable. Utah, as Mason noted, is a state where experience with state control of water pollution standards has not led to a lowering of standards. "Most people in Utah feel that we must have strong water pollution standards," stated Mason, noting how precious a commodity water is in his state.

A National Approach

Closely related to the issue of pollution standards and, like it, an issue that those who plan ways through which America's water infrastructure may be improved is the question of the necessity of having any type of "national approach."

"Streams don't stop at state boundaries," stated Swigart, making the point that even though America's water systems may indeed adhere to state and municipal boundaries, the nation's watersheds do not—a situation that indeed, some participants noted, does have an impact on the manner in which the infrastructure should be developed. This is particularly the case, participants stated, when water quality standards are involved. "What happens upstream is going to effect what those people downstream are going to do," one participant remarked, while another questioned whether downstream beneficiaries should help pay for improvements in effluent quality made upstream.

Regionalism—and regional concerns—were also raised by conference participants, and attention was given to the differences in needs, issues, and concerns of the Frostbelt versus the Sunbelt, and of water-rich versus water-poor states. Regional differences were also noted on an intrastate basis, with California being a prime example. "California has a unique system," Guess stated, since "it has no two identical water systems anywhere, nor an average water cost that makes any sense." Robert P. Miele of the Los Angeles County Sanitation District traced this variety in part to the state's geographic diversity. "Water is at a premium in southern California," he said, "and in northern California it is an entirely different situation."

While participants did not come to a consensus as to whether an approach to financing the nation's water infrastructure should be inherently "national" or not, they did raise the issue. Moreover, most probably agreed that the subject was one that did not lend itself to easy answers. As Guess noted, "Here is where we are going to find some very gray, hairy issues."

Funding Option

The papers prepared for the conference and reprinted as chapters in this book cogently present the views of conference participants on the various options and mechanisms that are available for rehabilitating America's water infrastructure, and no attempt shall be made to condense the content of those papers here. Nevertheless, the reader may find it helpful to learn of the types of funding options that attracted the most attention during the conference, and the intent of the following section is to provide such a preview to the more detailed analyses in the chapters that follow.

Outside of government grants—to be summarized later—the funding mechanisms that attracted the most attention by participants were user fees and bonds. "I think we can manage it with a user fee," Mason added. "We will, however, have to make some provisions for the needy." Others were also positive toward user fees, although a participant continued to question whether they might be relied on too heavily.

The widespread use by special districts of guaranteed water revenue bonds was pointed to as a possible solution. It was noted that the use of tax-exempt revenue bonds for a great variety of purposes had resulted in a very large volume of outstanding issues, and some participants raised the question of how far this could continue. However, the outlook here was rather optimistic. "Revenue bonds are replacing general obligation bonds," noted Peterson, a comment echoed by White, who stated that "some banking firms simply prefer revenue bonds to general obligation bonds."

It was noted that while a number of states have historically relied on general obligation local government bonds for water financing, the magnitude of the infrastructure-rebuilding situation may be such as to raise issues that might prevent such a continuation. Debt and tax limits and competing priorities were recognized as impediments.

Various other funding mechanisms were presented and discussed by participants. Savage discussed the possibilities (and problems) presented by permit fees and minibonds (tax-exempt bonds of small denominations for sale locally). White asserted that the problem could be solved by rate increases. "The most logical way to fund pure water is to give the provider adequate revenues," he argued; "we can do it without federal grants; . . . the key is adequate earnings and timely, adequate rate treatment."

Privatization

Privatization of U.S. water and sewage utilities, and thus of the renewal of the infrastructure effort, also received considerable attention. Early discussion centered on how much privatization was in progress, which, according to participants, was very little. "Most privatization schemes I've seen have dealt with wastewater," Guthrie stated, and although at least one example (from Suffolk, New York) of a municipality's attempting to sell its sewer system to the private sector was cited, very little privatization of sewage facilities seems to have taken place.

According to participants, the reasons for this are various, and a number doubted that privatization on a large scale was likely to occur. Peterson noted that the Internal Revenue Service, as well as a number of banks and lending institutions, might also be looking at privatization efforts in the water and sewage area with more than a little caution and that they are likely to urge the prospective buyer "not to be the first one in." White predicted that wastewater probably would not be privatized very rapidly without attractive income tax benefits for investors.

The less than active predictions for the privatization of America's water works failed to discourage a number of participants from voicing opposition to the principle of privatization. Much of the projected doubt about success for privatization, indeed, revolved around public policy issues. While one participant stated that the earliest water distribution systems were private, others pointed to a long heritage of public water and sewage systems. "Historically, water has been in our legal structure as a character of public," one participant argued. "At what point," another asked, "do you have the security of public health and welfare assured when [water and sewage services] are run by private industry?" Another questioned whether private industry had the ability to plan in the water and sewage area when "watershed planning has to be done on a state or a multistate basis." Another participant also questioned the effect of privatization on public input into plans for a renewal of the infrastructure: "With privatization you still have the public paying for a renewal of the infrastructure, but without having much to say about it." Not all the reasons why the public sector should not relinquish water and sewage services to the private sector, however, strictly involved issues of public policy. "The public sector," stated Torkelson, "should not give it [water] up just because it's profitable."

Participants also questioned the possible benefits that privatization might have—and the motives behind privatization efforts. "Where are they going to get the money for rebuilding the infrastructure?" a participant asked, and another questioned the notion that "the private sector can do it better." A number of participants also suggested that some privatization efforts were "tax gimmicks" aimed at providing tax shelters for the parent corporations—something that, according to one participant, "the IRS [Internal Revenue Service] will resist to the death."

Yet even with the generally negative reception that privatization received from the conference, some participants did identify areas where increased private sector involvement might develop. Primarily, these involved: the takeover of some small water and sewage companies; the provision of technical assistance in infrastructure renewal efforts, particularly to small utilities; and the privatization of some infrastructure repair efforts on a job-by-job basis.

The Role of Government in Renewing the Infrastructure

Notable attention was given by participants—both at the conference and in their papers—to government's role in rebuilding the water infrastructure: what it has been, what it should be, and when it is likely to occur.

The role of the federal government, not surprisingly, received the most attention—particularly the future of federal funding for water. While at least one participant declared himself opposed to federal grants, the majority of participants either voiced support for federal funds or refrained from voicing an opinion. One participant declared that federal monies should constitute 50 percent of all funds for infrastructure renewal, while another lauded the benefits of having federal funds in the enforcement of water standards through a carrot-and-stick mechanism.

A more discussed aspect of federal funding involved its continuance. "We're mesmerized by the gap and how to fill it," Peterson stated, "but I just don't see federal aid prospects that are encouraging in the long run." Torkelson also stated that he felt the federal government would not be able to fund the renewal alone, and that the federal construction grants program would not end but would change. The prospects of linking water projects with the federal jobs bill then being discussed by the administration and congressional leaders also received considerable attention from participants. One participant predicted that a jobs bill would focus on those areas where the federal government had the most capital invested (for example, highways), but he also stated that some funds for water might be forthcoming if one could match "short-term jobs with long-term products."

The possibility of receiving no federal funds for water infrastructure projects in the future—and its possible implications—was also discussed. "We've received funds from the federal trough for so long," a participant stated, "that it will be very difficult for us to get weaned overnight."

Participants also explored the relationship between federal funding and federal prescription, with most seeing them as inseparable. "He who has the gold makes the rules," White stated, and Savage asserted that "he who mandates should pay." According to some participants, Washington should receive considerably less than high marks in its current linking of funding and policymaking. "Congress is not putting its money where its mouth is," complained a participant.

Given the possibilities of a reduction or possible cessation of federal funds for water, participants also discussed the present and future of state and local government funding. According to Peterson, the "burden of financing is on the states, local governments, and private utilities," and Mason stated that "the bottom line is that we have to increase state and local dollars." Savage indicated that state officials are looking more closely at alternative funding arrangements. "State officials are just beginning to realize what funding options are available," she said, recommending that participants look at Wisconsin's "creative program" for ideas.

Ledbetter devoted a significant portion of his remarks to the various roles that should be played by the three tiers of government in the functioning of America's water systems. In brief, Ledbetter proposed that the federal role encompass the following: (1) research, (2) effluent and ambient standards, (3) ocean dumping, (4) interstate issues, (5) overview, and (6) technical assistance. He proposed that the states oversee the following: (1) waste load allocation, (2) issuance of permits, (3) water quality management, (4) compliance and enforcement, (5) ambient monitoring, (6) construction grants monitoring, (7) operator training, and (8) emergency response. Finally, he proposed that local governments oversee the following: (1) wastewater facility design, (2) construction, (3) operation and maintenance, (4) establishment of user fees, (5) pretreatment, and (6) effluent monitoring.

Time did not allow for a complete discussion of Ledbetter's proposals, which are further detailed in chapter 4 in this book.

Policy, Management, and Maintenance

Considerable attention was given by conference participants to certain elements of America's water systems that will need to be closely scrutinized and perhaps reworked in the years to come, regardless of what programs are adopted for rehabilitating the infrastructure.

Of particular concern was the way in which water issues are perceived and addressed. "We are reaching the end of an era," Peterson stated, "where clean water is considered as a single issue." For many participants, this translated into a future in which fiscal austerity and competing priorities will force water policymakers, planners, and managers to devise creative and more efficient ways of maintaining the current systems. "Because the financing gap is so large between what we can do and what we need to do," one participant stated, "we are going to have to close it at both ends. We need to reassess our needs, standards and priorities—and find more efficient ways of managing capital."

Others felt that more flexibility, particularly at the local level, on water policy and operations was needed. "Maybe we need greater flexibility at the local level to set goals," stated Mason, and Thomas issued a general call for "more local autonomy."

Participants urged water policymakers and managers to look carefully at their planning methodology. "We need to scrutinize our rules of thumb," Peterson stated, with other participants offering situations in which more scrutiny in the planning process was in order. Coelen gave an example where one community had allowed for some 2,000 years of growth in the design of its water system. Others debated the accuracy and nationwide applicability of life-of-pipe studies.

Participants also made specific proposals for improving the functions and operations of water systems. Mason suggested that industries be given incentives for pretreatment, and Guess suggested that these might come in the form of property tax relief for those plants that pretreat.

Technical assistance was cited by a number of participants as an extremely important element in improving water management—particularly in areas with small water systems. "In those areas where unsophisticated management still runs the system," one participant argued, "it will be very important to bring in a highly trained technical assistance service to help deal with operating and maintenance costs." Several participants urged that technical assistance be a state role, with some arguing for the creation of a special authority in each state to direct such programs.

Public Education and Politics

Public education was frequently cited as a crucial element in any agenda for rebuilding the nation's water infrastructure. Unfortunately, for many it was also seen as an area that presents very difficult problems.

According to a number of participants, the current outlook for governing public water projects is less than bright. Ledbetter identified six different factors that have helped to create such a climate: the establishment of unattainable goals; a negative public works image; an overkill of regulations; difficulty in identifying progress; oversimplification; and inconsistent enforcement. Other factors were added by conference participants, including, most importantly, the negative effects of the nation's current economic woes.

Many participants perceived the depth of public knowledge about the nation's water crisis as exceedingly low. "We educators know the frustrations of it," one remarked, "and I say this is one issue we'll never educate the public on." Others stated that most people are not aware that there is even a crisis in the infrastructure area and that educating the public had to be dealt with first. "The public doesn't know what the hell is going on," stated one participant.

Not all of the conference participants, however, were so gloomy about the prospects for adequate public awareness. Thomas took a more positive view, explaining how voters in Washington State have a high awareness of water issues.

The political dimensions of rebuilding the water infrastructure were also addressed by participants. While most appeared to agree with Guess that "first and foremost, water is politics," Savage's remark that "sewers and drainpipes don't make political platforms" brought up the sensitive— and natty—problem of making water infrastructure issues politically important. "Most political subdivisions are scared to death of these issues," one participant declared, and another termed them as "politically explosive."

Other participants made suggestions for ways to politicize water infrastructure issues. Michael stressed the need for water issues to be both easily identifiable and obviously important, such as having clean water or getting a sewer system repaired, while Deil S. Wright of the Department of Political Science at the University of North Carolina at Chapel Hill highlighted the desirability of a continuity of political leadership on water issues.

Where Do We Go from Here?

A final area discussed by conference participants was, not surprisingly, where do we go from here—a subject they approached with a sense of urgency. "The year 2000," Torkelson stated, "is rapidly catching up with history."

High on conference participants' list of needed future activities was the desirability of sorting out the nation's water infrastructure priorities. "The priorities should be set and followed in an order," Guess stated, adding that "we haven't yet got to the basics." Others concurred. "We've had so many ups and downs in policy that the messages have been thoroughly mixed," one participant stated, and another declared the need to "clarify what it is we are trying to do."

In order to facilitate this process, participants urged planning activities and the creation of test models. "We need a blueprint for the 2020s and 2030s," one participant declared, and others urged the Lincoln Institute of Land Policy to sponsor projects that test the fiscal and physical management of water treatment plants.

Whatever their specific suggestions, however, conference participants clearly relayed the message that a rebuilding of the country's water infrastructure was not only an important issue but one that this nation must address with deliberate speed. As one participant stated, "The time has come."

1

Clean Water:
The Financing Gap

George E. Peterson

The United States has suddenly discovered its neglect of the nation's public capital plant. Unfortunately, this discovery has occurred on the heels of a period, lasting almost two decades, during which the share of public investment in national product and in government budgets has been falling, and spending on capital maintenance measured in real terms has declined even more precipitously. The result is a capital backlog of vast proportions. Some observers have estimated that as much as $2 trillion, beyond current capital spending levels, would have to be spent over the next decade if all capital were to be restored to currently prevailing standards.[1]

A discussion of the prospects for financing this capital backlog should begin with two propositions. First, the country will have to reorder its spending priorities to generate more funds for public capital purposes. Second, it is as certain as projections about the future can be that $2 trillion of additional investment will not be raised. To raise it would require almost tripling the share of national product devoted to public investment—a change of budget allocation utterly without precedent, and one that would have to occur while the country is also trying to boost rates of private capital formation.

In my opinion, ultimately the infrastructure financing gap must be closed by generating additional capital investment *and* by redetermining what capital investment is truly needed. The public will have to look as assiduously for real economies in capital management as for federal capital aid programs. In addition, the country will have to systematically reconsider the standards it has set for capital facilities, whether these are built into federal law or incorporated into planning practice, and decide whether expensive standards should be preserved even if it means scrapping other types of capital projects altogether. At the same time the nation seeks to augment capital funding, it will also have to overhaul the institutions, political and economic, that have led to undermaintenance and postponed repairs in the past. It will do no good to mount a massive round of catchup investment if the deferral of routine maintenance and repair, which created the backlog in the first place, is allowed to persist.

Water and sewer systems are especially good illustrations of the dilemmas facing the country in making infrastructure choices. They are among the nation's most valuable public assets. They represent some of the oldest

13

pieces of capital, and in some parts of the country they give signs of nearing the end of their useful lives. The investment claims of water and wastewater systems have clear urgency, but in the aggregate they sum to far more than tax and ratepayers can be expected to pay. Selecting the right mix of additional capital commitment and reduced spending goals will be a difficult task—one made even more difficult by the apparent retreat of the federal government from funding support.

This chapter's discussion is generally restricted to the costs of funding clean water in urban areas. The cities are a paradigm of one part of the capital infrastructure financing problem. They have the oldest capital; often they have the poorest maintenance records; and frequently they face the largest repair and replacement backlog. However, the clean water problem is by no means an urban problem alone, and in some states it is the lesser part of the clean water financing dilemma.

Clean Water Capital Needs: An Overview

To place the estimates of capital needs for clean water in perspective, table 1-1 summarizes federally sponsored needs projections in five basic infrastructure functions: roads and highways, bridges, mass transit, sewer systems, and water systems.

It would be inappropriate to assign much precision to these estimates. They are very broad projections whose magnitude is highly sensitive to the particular capital standards adopted as reference points. These are, however, the hardest available figures on needs projections. Projections of a $3 trillion total backlog for public capital funding encompass a much greater variety of public facilities (ranging from canals and rivers to airports and from prisons to major dam projects), and the cost estimates are even less solidly grounded.

The cost estimates for water systems shown in table 1-1 refer to the capital costs for supply, treatment, and distribution in urban systems only. They were made in 1980 by the President's Intergovernmental Water Policy Task Force.[2] More than two-thirds of the estimated total of $75 to $110 billion capital investment is for the rehabilitation and replacement of existing distribution and treatment systems. Of this amount by far the greatest part is targeted for replacement of old water pipe.

The estimates of capital costs for wastewater systems come from the Environmental Protection Agency's needs studies. The greatest element of costs is found in the control of combined sewer overflows, though this also has been identified as one of the areas of less urgent concern to the federal government. After 1984 it will cease to be a categorically eligible use of federal wastewater grant funds.

Table 1-1
Federal Capital Needs Estimates Compared to Current Funding Levels
(billions of current dollars)

Infrastructure Category	Period	Coverage	Total Needs	FY 1980 Federal Outlays	FY 1981 Federal Authorizations	Total 1979 Capital Spending by All Levels of Government
Highways	1980–1995	Capital investment required to maintain existing highways at minimum condition standards. Does not include interstate completion costs.[b]	336.6 to 362.8[a]	8.7	8.9	15.7
Bridges	Current backlog	Capital investment required to rehabilitate or replace deficient bridges on and off the federal aid system.	41.1[c]	0.76	1.3	Included above
Sewerage	1980–2000	Capital investment required to meet water quality goals in existing federal law and to accommodate population growth.	19.0[d]	4.3	3.3	5.6
Transit	1982–1991	Capital investment required for rehabilitation of bus and rail systems and to accommodate projected passenger growth.	49.3[e]	2.3[f]	3.5[f]	1.6

Table 1-1 (continued)
(billions of current dollars)

Infrastructure Category	Period	Coverage	Total Needs	FY 1980 Federal Outlays	FY 1981 Federal Authorizations	Total 1979 Capital Spending by All Levels of Government
Water	1980–	Capital investment required to repair and replace urban water supply, treatment, and distribution facilities.	80.0 to 115.0	0	0	n/a

[a]Source: U.S. Department of Transportation, "The Status of the Nation's Highways: Conditions and Performance," January 1981. For a discussion of minimum condition standards (as determined by engineering performance and safety standards), see pp. 136–148. Highway capital needs reported here cover bridges on the federal-aid highway system.

[b]Interstate completion costs were estimated to be $50.75 billion in 1980. Source: U.S. Department of Transportation, "1981 Federal Highway Legislation: Program and Revenue Options," June 26, 1980.

[c]Source: U.S. Department of Transportation, "Second Annual Report to Congress, Highway Bridge Replacement and Rehabilitation Program," March 1981.

[d]Source: U.S. Environmental Protection Agency, 1980 Needs Survey: Cost Estimates for Construction of Publicly-Owned Wastewater Treatment Facilities, February 1981.

[e]Source: American Public Transit Association, Washington, D.C., preliminary results of 1981 capital needs study.

[f]Capital grants only, excludes any general operating subsidies used for repair.

Financing Sources

Where will the funds come from to finance these investments in clean water? The federal government is a shrinking source of funding. There is no federal financing for water distribution systems. The Corps of Engineers is not authorized to perform single-purpose functions such as the construction or rehabilitation of water supply systems. Proposals have been advanced to remove this restriction, allowing water supply facility construction to be a new purpose for federal water projects. Other proposed legislation would distribute all federal water resource dollars as a state entitlement, which could be used for various projects at a state's discretion.

In the aggregate, however, there seems little realistic prospect that the federal government will help bear the costs of renovating old water systems. Federal funding for wastewater facilities is on the wane. The 1981 Municipal Construction Grants Amendments cut federal spending for pollution abatement projects from $3.9 billion in 1981 to $2.4 billion in 1982. After October 1, 1984, the eligible uses of federal aid will be limited to wastewater treatment facilities, interceptors, and correction of infiltration inflow where cost effective. States may elect to use up to 20 percent of their grant funds on previously eligible projects such as sewer rehabilitation, new collectors, and control of combined sewer overflows. The federal share of project costs will drop from 75 percent to 55 percent, with the exceptions of previously approved projects and projects utilizing innovative and alternative technology.

The phase-out of federal capital aid for wastewater systems is consistent with at least part of the initial conception of the Municipal Construction Grants Program. When adopted in 1972, the program was not conceived as a permanent capital aid program but as a one-time effort to reverse worsening water pollution conditions. The design of the grants program reflected congressional intent of abating pollution on a national scale rather than providing financial aid to local wastewater systems. Although many of the original water pollution goals remain unattained, the federal government can point to a decade of unprecedented financial commitment to improving water quality.

There is some possibility that a public works jobs bill, if ever passed by Congress, would help the water and sewer systems. An evaluation of the last countercyclical public works bills found that approximately 23 percent of funds were spent on sewer and water system repairs. Any funding assistance from this source, however, will be temporary and skewed toward maintenance-type activities.

The long-run capital financing burden will fall squarely on state and local governments, private utilities, and most particularly on the local users of these systems. In principle, the capital for rebuilding sewer and water systems could come from general obligation bonds issued by localities and supported by general taxes. But, in practice, most of the financing will be raised by revenue bonds backed by utility rate changes.

User-fee financing provides the most stable revenue source for sewer and water utilities. Reliance on user financing to pay for these systems' capital requirements also is dictated by the entire infrastructure financing picture. Cities' ability to issue general obligation bonds has been weakened by taxpayer resistance to debt issues and by stiffer limitations on the volume of tax-supported debt. In the last four years the share of general obligation bonds in the tax-exempt market has fallen from two-thirds of total value to less than one-third. In their place have arisen revenue bonds as the mainstay of tax-exempt capital finance.

Certain classes of local capital projects, such as local road and bridge repairs and public-building construction, do not lend themselves to user pricing and cannot be financed through revenue bonds. These capital purposes have no alternative but to be financed largely through the issuance of general obligation bonds. From a city's overall capital financing perspective, it then becomes imperative that as many other capital purposes as possible be shifted to revenue bond financing to clear the general obligation market for these irreducible claims.

The scale of rate increases that will be necessary to finance the backlog of investment can be extreme. To pay for its negotiated agreements with EPA on secondary treatment facility investment, the city of Philadelphia has had to impose two separate rate increases of 40 percent each, in the space of eighteen months. It faces the prospect of further 13 to 25 percent increases annually over the next several years.

Institutional Changes to Support Financing

The main obstacles to placing water and sewer systems on a self-sustaining basis, capable of paying for their own capital requirements through revenue bonds, are not strictly financial but are also institutional. To gain market access for revenue bonds, a utility must have a secure revenue stream. That requires a pricing system which allows for recognition of capital and maintenance costs in the fee structure. It also requires assurance that rates will be raised in the future as is necessary to carry out system repairs and repay debt.

Moreover, for many government-owned utilities, especially those operated as part of the city's general budget, pricing policies have been prey to political considerations. Rate increases in those systems typically require the approval of city council. Rates generally have been held down, and approval of the rate increases necessary to finance capital spending have been far from automatic.

The inadequacy of their present rate structure, coupled with the uncertainty surrounding future approval of rate increases, has made it difficult for many urban systems to gain market acceptance for their debt. Sometimes the vulnerability of sewer and water capital financing has gone further. The cash reserves that utilities keep on hand for capital purposes have proved tempting to city governments plagued by their own budget problems. In the early 1970s, when the city of Buffalo first encountered financial problems, the city council passed a resolution requiring the water system to transfer some $2.6 million per year to the general fund. This action virtually eliminated capital repairs for the water system. When the city of Philadelphia faced a severe budget deficit in the mid-seventies, the city council simply declared that $20 million of the water system's cash reserve were "excess" reserves, and they also transferred them to the general fund.

Recognition of this institutional vulnerability has led a number of cities to consider divesting their sewer and water operations by setting up independent authorities. Such authorities are protected by law from transfers of funds to the general budget. They are endowed with independent rate-setting powers and can be required by their covenants of creation to establish rates that provide for adequate reinvestment and debt coverage. The purpose of such institutional reorganization is to create a market for water and sewer revenue bonds, thereby laying the financial basis for a long-term plan of capital reinvestment. Two brief histories, as follows, will illustrate the nature of such institutional reorganization.

Boston

Boston's sewer and water system was reorganized in 1977. Formerly run as part of general city operations, the sewer and water system had kept rates artificially low. By 1977 it faced a critical backlog of capital investment resulting from low capital spending and had an accumulated shortfall of some $25 million in its water and sewer receipts accounts. The system's rate of unaccounted for water was one of the highest in the nation.

With state authorization Boston set up an independent Water and Sewer Commission (BWSC), which has greatly strengthened both financing and capital reinvestment. By law it must annually draw up a three-year capital improvements plan with financing sources. In its first year of experience, the BWSC raised sewer rates 296 percent and water rates 16 percent. Unaccounted for water was reduced from 51 percent of total intake to 44 percent in the first three years of the reorganization. A long-term plan of revenue bond financing, secured by water and sewer rates, was put in place.

Cleveland

By the late 1970s Cleveland's water system had reached as advanced a state of deterioration as almost any water system in the country. The city's series of financial crises had diverted almost all maintenance and capital funds from the water system to the general budget. (This practice reached peak when the city, unable to find any other purchaser of its short-term notes, forced the water system to buy some $20 million of the city government's short-term debt; the city then defaulted on the notes). The city council kept the water rates among the lowest in the country. The financial shortfall was exacerbated by legal disputes between the city and its suburbs over which party had the obligation to carry out cleaning and maintenance of the city-owned water mains extending into suburban service areas. Neither party would take responsibility for maintenance for fear of establishing a precedent that it was legally bound to do so.

The suburban jurisdictions finally sued the city of Cleveland. A court ordered the city to surrender ownership of the water system to a new regional authority on the grounds that it was failing to maintain an asset of regional importance. Only with the threat of divestiture hanging over their heads did the parties negotiate a voluntary settlement, under which Cleveland agreed to carry out some $700 million of capital improvements to the water system. Annual water rate increases of 20 to 30 percent are projected over the foreseeable future.

Recapitalization of Old Assets

The decision to transfer a sewer or water system to a new authority represents a type of recapitalization. The city unburdens itself of its sewer or water debt and sometimes is paid an additional sum by the regional or other authority acquiring the assets. Central to the recapitalization is the ability to establish realistic rates for water or sewer services. It is this reorganization that converts the physical assets in the ground into an economic asset, capable of generating earning power.

In principle, recapitalization should be even more advantageous if an existing sewer or water system were sold to private buyers. Unlike public bodies, private purchasers can take advantage of the federal tax depreciation laws.

In what promises to be a landmark case, Suffolk County (New York) is trying to do just that with its Southeast Sewer District. It has structured a sale and lease-back agreement whose net effect is to transfer title to the sewer district to a consortium of private investors. In return for transferring the depreciation benefits, the county will receive a cash payment of roughly

$180 million. The size of this deal is such that, if finally approved by the IRS, it promises to trigger a large volume of similar sale and lease-back arrangements.

Reducing Capital Investment Requirements

At times the policy debate over capital rebuilding seems almost mesmerized by the search for new financing sources. But, in the end, the capital financing gap will have to be closed by scaling back investment requirements as well as by generating new funds. Few segments of the public capital plant better illustrate the type of decisions that will have to be made and the opportunities for cost saving that exist than water and sewer systems.

Water Pipe Replacement Policy

In all projections of the capital costs for urban water systems, the largest investment category is the cost of replacing old pipe in the distribution system. The Presidential Task Force study, cited earlier, assumed that over the next twenty years all pipe reaching seventy years of age would have to be replaced. Given the age of the water distribution systems in most northern and eastern states, this rule of thumb automatically generates a large volume of capital spending.

Several individual cities have made similar projections of their replacement needs. A special study prepared for the city of Buffalo assumed that all pipe would be replaced at sixty years of age. This policy would have required a manyfold increase in the city's capital spending for the water system. The New York City planning commission, for a time, based its estimates of capital spending requirements on a seventy year useful life for water pipe. Boston's capital plan projects a ninety- to a hundred-year life.

There are two noteworthy facts about these projections. First, given the age of U.S. water distribution systems, the difference in the capital spending dictated by a sixty-year replacement rule and a one-hundred year replacement rule can be very great. Second, from all the available evidence, replacement of water pipe before failure, based on age alone, is an unconscionably expensive strategy. Studies by the Corps of Engineers in Manhattan and the Environmental Protection Agency in Cincinnati have demonstrated that the probability of water main breaks is only loosely related to the age of pipe segments.[3] In the Manhattan study, it was estimated that the same reduction in main breaks could be achieved at one-third the cost if a replacement strategy based on the actual probability of pipe failure was adopted.

In fact, there is not a water system in the country that, under the constraints of its own capital budget, actually pursues a policy of replacing all pipe once it reaches a certain age. Yet calculations of the cost of doing so dominate estimates of national and local capital "needs" for urban water systems. The urgency of replacing overage pipe has become the lynchpin in congressional proposals to devise a new federal funding policy that will assist older states in the cost of water system replacement. Perhaps it is unfair to take these arguments too literally, but a national infrastructure policy that actually compensated states and localities for the cost of replacing water pipe, based on age alone, would be an extravagant misallocation of scarce capital resources.

Wastewater Standards

The magnitude of capital investment at stake in establishing infrastructure standards is well illustrated by the current consideration being given by the EPA to applications for waivers to permit ocean discharges. The 1981 construction grant amendments allow certain coastal cities to apply for waivers of the full secondary treatment requirements if the city can demonstrate that there will be no adverse effects to area waters. New York City has estimated that a waiver on its new North River treatment plant, to allow the city to complete only the first phase of the treatment plant which is capable of removing 25–35 percent of standard pollutants, would save $250 million in capital costs. The General Accounting Office (GAO) has estimated that with broadened eligibility for ocean discharge permits, as many as eight hundred waiver applications are possible.[4]

Management of Risk

One promising avenue for reducing capital costs is the explicit assessment and management of risk.

For most of this century urban water managers have constructed their systems to ensure undiminished water supply for at least 95 percent of the summer droughts that a region can anticipate. Projected shortfalls in water supply have been met by developing new sources of supply and new reservoirs.

A recently completed agreement by the Washington Suburban Sanitary Commission and neighboring water districts illustrates the capital cost reductions possible from explicit demand and risk management.[5] In this case a capital plan that would have required a $1,400 million investment was reduced to roughly $30 million in capital outlay. The reductions in capital costs were made possible by the following policies:

An increase in water rates and a shift to increasing block-rate pricing lowered demand for water use, thereby eliminating the need for part of the planned addition to capacity.

An explicit shift from a 95 percent safe yield standard to a 92 percent safe yield standard was judged acceptable because it could be absorbed solely by outdoor lawn and garden watering. A two-year involvement of area political leaders and citizen groups culminated in political endorsement by all participating jurisdictions of the reduction in safe yield and the water management policies that would be followed during drought periods. This further reduced capacity requirements.

The reservoir system was redesigned so that instead of relying on greatly expanded reservoir capacity at the source of the water, much smaller additions to capacity were made nearer to the location of use. In this way, releases of water into the Potomac River could be made one day ahead of consumption, when more exact information on the daily volume of needed release was available. Otherwise, water in much greater volume would have had to be released from reservoirs three days upstream. This too reduced capital requirements.

Conclusion

Other chapters in this volume are devoted to the specifics of financing water and wastewater systems. It seems increasingly likely, however, that the broad choices about capital investment in these systems will be made against the backdrop of more comprehensive policies toward infrastructure reinvestment. From this perspective certain conclusions are relevant.

The total cost of the public capital investment backlog, as currently defined, seems well beyond the country's ability to finance. Difficult choices about capital priorities, therefore, will have to be made. The impetus of federal legislation, as well as states' capital financing policies, is to move toward unified capital budgets where broadly competing capital claims must be weighed against one another.

The prospect that there will be a significant shortfall in capital financing will place a premium on efforts to lower capital costs through more efficient management. The capital financing gap must be closed from both ends—by generating more resources and by reducing capital investment requirements—not by a search for funding sources alone.

The final cost of paying for clean water will be borne primarily by system users, both because user financing provides the most stable basis for capital investment and because general obligation bond financing and general tax sources must be reserved for those capital functions that are unable to charge fees for their services.

Widespread conversion to full capital costing in the rate structures of publicly owned utilities will often necessitate basic institutional reforms in the ownership and management of capital. This will be true especially in the older cities. The need to reorganize institutions in order to issue revenue bonds will prove to be the entering wedge for more lasting reforms in the way public capital is managed.

Notes

1. The Associated General Contractors of America, *America's Infrastructure: A Plan to Rebuild* (Washington, D.C., 1983).

2. *Urban Water Systems: Problems and Alternative Approaches to Solutions* (The President's Intergovernmental Water Policy Task Force, June 1980).

3. Robert M. Clark, et al., *Determinants and Options for Water,* Betz, Converse, Murdock, Inc., Corps of Engineers, *New York City Western Supply Infrastructure Study* (Plymat Meeting, Pa., 1980).

4. General Accounting Office, *Billions Could Be Saved through Waivers for Coastal Wastewater Treatment Plants* (Washington, D.C., 1981).

5. Robert J. McGarry, "Potomac River Basin Cooperation: A Success Story," paper prepared for National Research Council Conference, *Cooperation in Urban Water Management* (October 1982).

2

Funding Clean Water: A Private-Sector Perspective

John W.L. White

This chapter describes briefly the public water supply industry in this country and where my company, Consumers Water Company, fits into the industry.

The EPA reports that there are some 61,000 "community water systems" in the United States. This figure includes a huge number of very, very small unregulated systems, such as small mobile home parks.

The overall water supply industry is represented by the American Water Works Association (AWWA), which has 27,000 individual members and both municipalities and investor-owned companies. It establishes standards for materials and equipment and recommends operating policies that are adopted throughout this country and the world. In addition, the investor-owned side of the industry is represented by the National Association of Water Companies (NAWC).

It may help in visualizing the investor-owned water industry to know that there are approximately 6,000 "regulated" investor-owned water companies in the United States, serving approximately 35 million Americans in thirty-two states with more than 2 billion gallons of water per day, with annual revenues totaling about $1 billion. Perhaps 90 percent of these companies are small developer systems that were created to serve new housing developments. There are perhaps two dozen investor-owned systems whose stock is publicly traded with a regular market—in most cases over the counter—and many medium-sized companies not publicly traded.

The most important role of the investor-owned water industry was well described in a report submitted by Day & Zimmerman Consulting Services for presentation to the National Water Commission in 1972. This study, which detailed the case histories of six investor-owned water companies, stated, "As private corporate entities, the regulated investor-owned water companies can own property and serve customers in all jurisdictions within specified geographical areas. They are characterized by having continuity of utility management, fast decision-making potential, sound business and financial policies, flexibility and adaptability in meeting national urban water goals, and elimination of undesirable political pressures on utility service. Another important advantage to the public is that the regulated investor-owned water companies pay local, state and federal taxes, as do other private businesses."[1]

Consumers Water Company

Consumers Water Company—the company with which I have been associated for over thirty-five years—is perhaps the tenth largest in the industry. Consumers was founded in 1926 and now serves approximately 150,000 customers located in seventy-eight communities in seven states. It has revenues of approximately $35 million and gross plant of over $100 million. Its stock is traded on the over-the-counter market, and there are approximately 3,500 common shareholders. Consumers also serves sewer customers in Illinois, owns a real estate company, and is involved in a very limited amount of gas-drilling activities in Ohio.

The company has a professional organization made up of engineers, accountants, and business managers. It is currently exploring the possibility of offering some type of management-consulting services to municipal and quasi-municipal water and wastewater systems. Recently the company, together with an experienced well driller, made a joint-venture proposal to a municipality in New England to locate and develop an additional water supply for that city and to sell the city water on a long-term contractual basis.

Need for Adequate Earnings

The funding of pure water is a very important and timely issue. When the question of how to fund public water supplies is raised, the answer is, in my opinion, relatively simple in most cases. Simply stated, "With adequate earnings, anything can be done!" In other words, the most logical and equitable method of funding pure water is to assure that the supplying entity is receiving sufficient water revenues to be a financially sound and viable business enterprise. If the entity is financially sound, then it will have the financing capacity to sell appropriate securities on competitive terms to build or rehabilitate whatever type of water system is needed to meet the particular requirements of the geographical area served, without the need of any federal grants.

The typical capital structure of an investor-owned water company would be 35 percent common equity, 5–15 percent preferred stock, and 60–65 percent debt. Fair and timely increases by the state public utility commissions and/or the municipal entities served are absolutely necessary to meet the required financial coverages. To repeat, the essential ingredient is sufficient water revenues received in a timely manner to meet the costs of attracting the appropriate capital structure. If this is accomplished, no federal grants are needed.

The underlying ingredients necessary for funding clean water continue to reappear in the study of successful entities. Some of the more significant are fiscal responsibility, innovation, mutual trust, cooperation, understanding, teamwork, professional management, political courage, and fair and timely regulation—all of which will be quite apparent in this chapter's discussion and specific examples.

With reasonable adherence to these simple qualities, the problem of funding clean water becomes both solvable and a source of personal satisfaction to those involved. Without those ingredients, efforts to provide quality water service become extremely frustrated, which results in poor service to the customer.

Government Grants

According to the AWWA, "Governmental grants to aid water utilities in the construction of necessary facilities are undesirable in that they destroy the financial and managerial independence necessary to self-sustaining businesslike operations."[2] This is a position I support.

Some ten to fifteen years ago there were substantial federal grants available to various municipal water systems, regardless of the financial need of the entity involved. These grants were very distressing to many in the investor-owned field since municipal officials and customers frequently asked why the private water company rates were higher than those of neighboring municipal operations. Most of these municipal operations paid no federal, state, and local property or income taxes, and yet some were receiving substantial grants from the federal government—grants that were funded in part by the income taxes paid by investor-owned companies. This was hardly an equitable situation.

One brief example will demonstrate the ridiculousness of some of these grants. A large and well-run quasi-municipal entity in Maine received more than $750,000 to install a major feeder main. This entity enjoyed at least an AA financial rating and had absolutely no need of a federal grant. The general manager of this regional water system recognized the inequities of a municipal water system's receiving government grants when such grants were not available to investor-owned systems. He stated to me that he did not believe in the grant program but that his responsibilities were to manage the water system and that, since the federal laws made grants available to municipal systems, it was incumbent upon him to apply for them. I reiterate that this entity had absolutely no financial need for a grant. It could easily have borrowed the money and adjusted rates to reflect the added capital cost.

Federal leadership is often called on for new programs because of the expectation that with such leadership will come federal funds. However, there is a new version of the age-old, time-honored Golden Rule that states, "The one who has the gold makes the rules," and many persons who have had experience with federal construction grant programs for wastewater disposal systems now believe the price of federal intervention may be too high.

It is hoped that this type of federal grant will not be available in the future—although currently there is considerable lobbying in Congress by many who are in favor of such grants. The Comptroller General of the United States, in a report to Congress on urban water distribution systems (1980), stated, "A move is on to modify federal policy to provide more aid to help large cities rehabilitate and improve their water supply and distribution systems. Such programs, if enacted, could add billions of dollars to the federal budget. The General Accounting Office (GAO) believes that the case for more federal aid for urban water distribution systems is not convincing, and that legislation to provide such aid should not be enacted until the needs are clearly established."[3]

This GAO report, analyzed in detail three urban water systems: Boston, New Orleans, and Washington. Three quotes from this report are relevant:

> GAO believes that if, as the subcommittee reported, most water systems are financially self-sustaining, there should be little need for federal aid.

> Older water distribution systems are often pictured as antiquated, poorly functioning, inadequately maintained, and on the verge of complete collapse. To test the validity of this dismal impression, we performed extensive work in the three cities just mentioned. The Boston distribution system is often depicted by the news media and in other forums as aged, poorly maintained, and in danger of immediate collapse. However, Boston officials believe the system was generally in good shape, and they conveyed no sense of impending doom in our discussions with them.

> Municipally-owned and operated systems are four times as likely as privately-owned systems to experience shortfalls [in revenue dollars].

Comparisons

It is generally accepted in the water utility business that a meaningful comparison of water rates between water suppliers is nearly impossible without at least a detailed analysis of many factors. Nevertheless, the general public frequently makes comparisons without such analysis. The fact is that each water system is unique unto itself, and the cost of providing service depends on the unique features related to that particular system. For example, Consumers Water Company's systems has approximately thirty-three different rate schedules for its various companies. For an average family using approximately 1,600 cubic feet per quarter (about 133 gallons per day), the rates run from a high of perhaps $60–$65 per quarter (three months) charged by small systems with water supply difficulties, down to about $25 a quarter charged by a large system sitting on the edge of Lake Erie. Seven primary factors bring about the rate variances between its own companies, as follows:

1. *Difficulty in obtaining a supply.* As nearby water supplies are used up, supplies must be obtained from greater distances.
2. *The customer density.*
3. *Investment per customer.* This amount is often affected by how long ago the system was built or expanded and the level of construction costs and/or interest rates at that particular time.
4. *Treatment costs.* This amount is significant particularly where water softening is necessary and where some of the more sophisticated requirements of the EPA are involved.
5. *Level of property taxation.*
6. *Wage rates.*
7. *The percentage of total revenue absorbed by the municipality through fire protection charges.*

A proper comparison of water rates requires a detailed analysis of these factors to determine why rates are higher in one community than in another. As a general rule, the economies of scale will dictate that a small company with low customer density and a difficult supply problem will result in higher rates. In comparing investor-owned water system rates with those of municipal systems, the primary variables are those of taxes and the degree of federal grants and/or FMHA subsidized interest rates. On the average, investor-owned systems pay from 15 to 20 percent of their revenues in various forms of taxes.

Cost of Water as a Proportion of the Family Budget

In focusing on the subject of funding pure water, it is of interest to compare the annual cost of water service with the cost of other items in the family budget that many people consider necessities of life. In the past water has been so cheap in most communities that to obtain an increase of, say, 25¢ per day resulted in a very substantial percentage increase. This dilemma often caused municipal officials to postpone needed increases in order to avoid what they feared might be an adverse political reaction from the voters.

Water costs need to be placed in perspective: Not only is water the single most important element of survival, it is one of the least costly items in the family budget. It must be recognized, however, that in future years this very low cost of water service may become a thing of the past. In fact, any charge less than $1 per day should be considered a bargain! Consumers should look at the price of water on a monthly basis (or preferably on a daily basis) rather than on a three-month basis, as it is currently billed by most water purveyors. Many water users compare their quarterly water bill with the *monthly* telephone, gas, or electric bill and tend to forget that the water bill is usually providing service for a three-month period.

Although a comparison with other items in the family budget may not be a particularly scientific approach to setting water rates, it does, nevertheless, put into perspective the value of water service in comparison with other regular expenditures, many of which are paid for on a daily or weekly basis, causing one to overlook their true burden on the pocketbook.

In 1973 I was a trustee on a small sewer district in Freeport, Maine, that was undertaking a capital expenditure program of between $3.5 and $4 million, most of which was paid for by the federal and state governments. The proposed sewer rates were to cover primarily operating expenses (with very little capital cost) but, nevertheless, would result in very large increases over prior rates.

In an effort to put the proposed rates—which we estimated might be over $100 per year per customer—into perspective and also to have some fun in the process, we prepared a short questionnaire to be filled in by the trustees and others in management. Each estimated what they were paying on an annual basis for six other family expenses. The results were quite interesting: compared to electricity, telephone, cigarettes, heating oil, milk, and gasoline, water was by far the least expensive.

In a somewhat similar vein, one of our companies in Maine—Camden and Rockland Water Company—compared its average water cost per day of 30¢, or $108 per year, to nine other commonly purchased items in the family budget. Again the water service was the lowest item.

The presentation of a comparison of family budget costs as just described is in fact helpful in adopting the basic philosophy that water rates should and can be self-supporting. It is common knowledge that percentage increases can be *very* misleading. A new water supply or the rehabilitation of old mains may well require a 50 percent or 100 percent increase in water rates in extreme circumstances and still be significantly less than $1 per day. In most cases such increases *can* be absorbed in a family budget. The mere necessity for the substantial percentage increase should not bring about acquiescence to political pressure for federal assistance.

One final note on this subject: $1.00 to $1.25 per gallon is the approximate price of drinking water in most supermarkets in Saint Croix, U.S. Virgin Islands, and in Key West; rates charged by the tax exempt public authority average $50 a month for a single-family dwelling. Both examples give an economic indication of the real value of pure water when it is scarce.

Revenue-Supported Financing

Four examples, as follow, of innovative revenue-supported financing techniques that have been used or at least considered in recent years to expand the availability of quality water service by investor-owned systems are relevant to this discussion.

Garden State Water Company, New Jersey

The Garden State Water Company, owned by my company, was created by the consolidation of about ten fairly small, troubled water systems. It currently serves three primary areas in New Jersey: Phillipsburg and environs; Hamilton Square which is outside of Trenton; and Blackwood which is outside of Camden. There are approximately 80 miles between the two systems farthest apart. Consumers' first entry into New Jersey was in 1964 with the purchase of the Blackwood Company. This sleepy little town had felt the effects of rapid expansion following World War II and had grown from about 400 customers to approximately 1,200 customers by the early 1960s. In 1965 the company acquired the Hamilton Square Water Company serving approximately 3,000 customers; four years later it purchased the Phillipsburg system, serving about 6,000 customers. In addition, it acquired six other small mostly troubled water companies combining them into a newly created Garden State Water Company, which presently serves twelve different government entities. During this eighteen-year period the number of customers has increased—through acquisition and growth—from 1,200 to almost 20,000, with a gross plant of $18 million.

Through innovative procedures, a spirit of cooperation and understanding with government officials, the company has grown and it now has professional management, reasonable earnings, good credit, and is a recognized leader in the industry. The combined entity is large enough to economically attract needed debt capital that would have been an impossibility for most of the original entities that Consumers' had acquired. The key to financing this growth situation was adequate earnings coupled with innovation, understanding, teamwork, and trust.

Freeport, Maine

Historically, Freeport has been principally a shoe shop town; it is also the home of L.L. Bean's Mail Order Service. The town was served by the Freeport Water Company and by two small water districts—one serving less than 50 customers and another about 150—located some distance from the primary Freeport system. Freeport Water Company purchased the assets of the smaller of the two districts and terminated its existence. Consumers' interconnection with the other two districts has permitted the two systems to operate almost as one. The Freeport superintendent and his assistant manage the South Freeport District under the direction of an elected board of trustees.

In 1972 a potentially disastrous fire caused the town to focus on its need for a substantial additional water distribution storage and main capacity. Working with a citizen study committee, the town financed an 800,000-gallon stand-pipe and the water company installed additional mains of 12-inch and 16-inch size. The town still owns the tank and leases it to the company for $1,000 a year. This cooperation permitted the building of the tank at the lowest annual carrying cost to the ratepayer.

Two years ago extremely dry conditions and increased water usage in the area caused both the district and the water company to *require* additional water supply capacity. Test drilling was undertaken over an extended period of time with relatively little success. Rainfall was substantially below normal and the groundwater levels dropped, which resulted in an emergency condition. The solution was the installation of 1½ miles of 8-inch mains and a pump installation that tied the South Freeport Water District mains in with another water district in a neighboring town to allow purchase of additional water when needed. The project was put into service in January 1981. Of major significance was the degree of cooperation by four different entities involved—first the town of Freeport agreed to contribute $50,000; then the Freeport Division of the Maine Water Company and the South Freeport Water District agreed to split the remaining costs; and finally, the Yarmouth Water District agreed to the interconnection. *No Federal Funding was involved.* This is a good example of the tenet that with adequate information, understanding, and cooperation, the public—in most cases—is willing to pay whatever it costs to fund clean water. What the public does not want is poor quality, or to be told there is no water available and that they cannot use the amount they feel is necessary to meet the needs of their families.

Riverton Consolidated Water Company, Pennsylvania

The third example is an innovative financing program now in the process of development by the Riverton Consolidated Water Company (a subsidiary of American Water Works Company), which serves approximately 25,000 customers in a middle- to upper-class suburban area immediately west of Harrisburg, Pennsylvania. The territory is primarily residential and has a great deal of potential growth. This growth, however, is partially limited by the unavailability of either public water or other acceptable private sources. To effectively serve the area, the water system will require a $3 million transmission loop plus additional distribution mains in order to serve about 3,000 building lots. The three parties involved were the township, some builders, and the water company. All were anxious to work out a solution. A proposed solution, which is still in its infancy, is as follows:

1. The township would finance the initial project.
2. The builders would pay $1,000 per lot sold to retire the initial debt.
3. The water company would pay the interest on the debt, which in return would be capitalized with the interest payments being treated as part of the rate base.

The advantage to the township is that it would grow and gain tax base; the builders are able to develop 3,000 new homesites; and the water company will be able to use more of its existing plant capacity. The company reports that joint cooperation so far has been outstanding, and it is enthusiastic about being able to resolve any difficulties.

Hackensack Water Company, New Jersey

This company was undertaking a major expansion, which it hopes to complete in 1986, known as the Wanaque South Project. This project consists of enlarging an existing reservoir, constructing a 250-MGD (million gallon per day) pumping station, and installing a 17-mile aqueduct 5 feet in diameter, and expanding a filtration plant at a total estimated cost of some $122 million. The program will double the company's plant investment by 1986.

Hackensack is one of the largest investor-owned companies in the country, serving over 213,000 customers in sixty communities in northern New Jersey (plus Rockland County, New York), with total operating revenues for 1981 of over $62 million. The undertaking of this project, even in the best of financial conditions, would be immense; however, Hackensack has been through an extreme drought period coupled with statewide water restrictions that resulted in the company's experiencing an actual *net loss* of $499,000 for the first six months of 1981. During the drought, the service and public relations problems were horrendous, and over 21,000 customers paid state-mandated penalties for exceeding water-use allotments during the nine months of 1980 to 1981 when Hackensack's service area was under emergency rationing.

The financing of this huge project is being carried out to a large extent by the company's issuing—through the New Jersey Economic Development Authority—$100 million of tax-exempt bonds that have been insured, both for principal and interest, through an agreement with the American Municipal Bond Assurance Corporation (AMBAC), resulting in a AAA bond rating. This unique feature—and probably a first for water utility bonds—provided a significant savings in interest rates and in turn costs to the water customers. The New Jersey Board of Public Utilities was most cooperative in their allowance of adequate rates to carry out this project, including authorization to adjust water rates and include construction work in progress in the rate base on a quarterly basis. This innovative rate treatment permitted the company to increase its dividends and placed the company in a position to sell additional common stock to provide a portion of the equity requirement of the project. The requirement of the bond financings were that the level of total debt including short term must not exceed 70 percent of the capitalization. This project could not possibly have been financed without much cooperation between state and local authorities, the financial community, and the company management striving together to create an innovative solution to a very difficult problem.

It is ironic that the company had been stressing the importance of undertaking this project many years before the recent drought, but due to political problems, the necessary approvals were not obtainable until this recent situation created an almost unprecedented emergency. The key to solving the large and unique financing of this project was the timely, adequate, and innovative rate treatment, which gave adequate assurances to the financial community. As a result of this realistic rate treatment, the company quickly went from a loss situation to a sufficient profit level to permit its stock to be selling above book value.

Conclusions

In an effort to offer specific suggestions on the subject of funding clean water, I have sought commentary from the presidents of my company's subsidiaries, members of the executive committee of the National Association of Water Companies, and others with many years of experience in the business. The following is a list of policies, procedures, and/or techniques that the investor-owned segment of the industry believes to be important in addressing this most important subject.

Responsible regulatory treatment, primarily adequate and timely rate relief (with negotiated settlements being encouraged) that will permit the utility to have a reasonable opportunity of achieving the return on equity allowed and promote the development of incentives to encourage well-run and efficient systems.

Develop innovative rate techniques, such as single tariff pricing offset charges, pass-through items, step increases, and a forward-looking test year to minimize attrition in recognition that water utilities are probably the most capital intensive of all businesses.

Self-supporting main extension policies, including "back-up" facility charges, thus eliminating the need for the existing customers to subsidize the increased capital costs of serving new customers.

Eliminate the so-called free rider concept by requiring new customers attaching to recently installed mains to pay a proportional cost of that extension.

Tax-exempt financing: Initiate statutes, if necessary, to encourage the further use of tax-exempt financing together with guarantees, prepayment, or "redemption bond" provisions, and other innovative features that will reduce the cost of borrowing money.

Discourage proliferation of developer systems: adopt state and/or local legislation or health regulations that will discourage the proliferation of developer water systems (or individual private wells) when interconnection to a larger well-run central water system is feasible.

Fair taxation of watershed land: adopt state statutes requiring that watershed land be taxed no higher than comparable woodland property.

Appropriate amendments to Federal Safe Drinking Water Act by adopting HR-4509, the Safe Drinking Water Regulatory Reform Act, applying more reasonable standards under which EPA could prescribe treatment techniques thus giving more flexibility to the water utilities and states.

Reduce problem of small troubled water companies: provide economic incentives that would encourage the larger well-run systems to acquire the small troubled water systems, most of which were originally developer systems.

Equalize financial treatment by the federal and state governments, for both municipal and investor-owned systems, on such matters as water main relocations resulting from highway construction and equalize terms of revolving long-term loans at competitive interest rates to those water systems that are unable to obtain their own financing.

Continue the utility dividend-reinvestment program: The very successful plans permitting federal tax deferral of dividends scheduled to
expire in 1986 should be continued.

Encourage joint public investor-owned projects to assure improved
regionalized service through interconnections and lower costs from
economies of scale.

Encourage greater understanding by the public of the relative low cost
of water as compared to other items in the family budget and the importance of meeting water needs without looking to the federal government.

Encourage conservation and appropriate pricing, particularly for large
users, to discourage waste, thus keeping down the costs of financing
unneeded capital expenditures.

In closing, I would like to share a different approach that is rather
unusual but quite complementary to the investor-owned water industry.
Steve H. Hanke, senior economist with the Reagan administration Council
of Economic Advisors, suggests that to solve our urban water problems, we
should move rapidly toward the "privatization" and decontrol of the water
industry and away from its politicalization.[4] Whether the evolution goes as
far as Hanke suggests, or whether it basically results in closer cooperation
between the entities remains to be seen. As our company, however, has continued its investigation of possible income tax benefits that could result
from the privatization concept, we have concluded that there may be applications for its use that could be quite attractive to all concerned. We look
forward to becoming involved in such a project.

Notes

1. Day & Zimmerman Consulting Services, "Publicly Regulated Investor-Owned Water Utilities," Report to the National Association of Water
Companies, January 21, 1972.
    2. American Water Works Association, Board of Directors, "Organi-
zation and Management of Publicly Owned Water Utilities," policy statement adopted 1970; reaffirmed 1977, rev. 1981.
    3. Comptroller General of the United States Report to the Congress,
"Additional Federal Aid for Water Distribution Systems Should Wait Until
Needs are Clearly Established," November 24, 1980.
    4. *Wall Street Journal,* September 3, 1981.

3

Government Subsidies for Clean Water: Who Wins and Who Loses?

Stephen P. Coelen and
William J. Carroll

Water supply in the United States is a critical resource, valued for sustaining life itself. As such, we have protected the supply as a treasure, not always conserving it as we should but increasingly becoming concerned about its quality. Since the late 1960s, and certainly from 1972 when the Water Pollution Control Act was implemented, the federal government has sought to perpetuate a continuing source of pure water. Most of the early legislative acts that dealt with water supply were aimed at regulating the quality of outflow from sewer systems. The National Rural Survey and discoveries of the existence of major contaminants in urban water supplies led to passage of the Safe Drinking Water Act, the guardian of the nation's water supply. Between the two acts, tremendous funds were generated for supplying wastewater and water supply treatment facilities. The EPA generated a wastewater treatment facilities needs survey every biennium in the 1970s. Two water supply inventories were generated during the same period.

Funds for wastewater facilities have been tied to future growth by the EPA "five-fourths' rule" that limited the size of facilities that could be developed based on five-fourths of the area's 1972 population. Funds for water supply facilities have been tied to future growth in a similar fashion. The clean drinking water act provided funds for twenty years' growth, a provision that was only recently withdrawn. Further, when added together, the implicit population projections developed by local areas for grant applications for wastewater sewer facility development suggested a doubling of the population in the country from 1975 to 1990—an obvious impossibility. This became a major force in the Office of Federal Statistical Policy and Standards directive that the Bureaus of Economic Analysis and Census jointly develop population projections for small areas so that they would be consistent with total national growth scenarios. At the same time Urban Systems, Inc., conducted research on the use of interceptor sewer construction funded by the EPA, determining that the construction would frequently satisfy hundreds of years' worth of growth, should growth continue at historical rates.[1]

Federal involvement in clean water has been invoked on the basis that individuals and communities were too unaware of the microconditions of their water and effluent to know that it was unsafe. Further, and more convincingly, pollution emitted in one area flowed frequently enough across state borders that federal involvement was deemed necessary. At the same time several states became involved because of the insolvency of some of their local water supply providers. This typically occurred when investor/developer-owned utilities were taken over by the communities they served. Finding capital equipment unrepaired and fully depreciated and operating without trained supply and treatment operators and lacking other options, many such systems declared bankruptcy and asked for state intervention.

Of course, not all the states' activity was prompted by emergency situations. Many, including North and South Carolina, Washington, New York, and California, were involved early, looking for creative ways of financially assisting communities and at the same time infusing them with technical assistance and concern for capital maintenance programs.[2] New York, for example, specifically implemented an operations and maintenance subsidy for systems that was designed to help systems maintain and protect investments otherwise often mistreated. The bulk of such efforts, however, was aimed at providing assistance for funding the capital requirements of utilities and ensuring that operators were sufficiently trained for required duty. Even the New York operations and maintenance subsidy was not renewed after its first legislative success.

In summary, the bulk of financial activity from state and federal governments has been the provision of necessary capital for construction of facilities, the regulation of quality by setting standards of maxima for contaminants, and the permissiveness to allow capital grant applications to seek funding for future population growth. This in many cases has led water systems to overcapitalization—a phenomenon not uncommon in public utilities. Overcapitalization has been the subject of many hypotheses arguing that rate-of-return regulation gives rise to conditions under which utilities extend their capital rate base in order to earn a greater total return.[3] Such overcapitalization has led to the provision of capacity for fringe populations, that is, those outside the original service area, because larger populations reduce average costs. Local communities may be urged to see this as helpful whenever capital, fixed-cost spending is enormous as it is compared to operating-costs. Finally, overcapitalization has led to geographically specific, "sprawlish" effects since the financial returns to the extension of water supply/sewerage service are larger whenever the recipient land areas are larger than minimum size.

These observations are the subject of the following sections in which emphasis is directed toward analyzing the distributional impacts of investments.

Economic Considerations of Subsidies

Whenever there is involvement of state or federal governments in assisting water utilities with their compliance with water quality standards, there is a prima facie case that subsidization is taking place. Warford, writing about the appropriateness of various water prices in the South Atcham Scheme in England, suggests that this principle also applies to a variety of service functions.[4] In their study they recognize that water is not the only resource endowed with such subsidies. Road construction, electrical supply, telecommunications, and other public investments all carry intraregional subsidies. Once a country begins such a program, in order to maintain intraregional equity, further requests for similar subsidies cannot be turned down.

The regional capture of higher-than-average subsidies across a variety of investments carries with it a systematic rationale that individuals will be biased in locating in one place rather than another. This precis is at the very foundation of arguments that subsidies cause inefficient distribution of regional activity. Activities tend to locate in a particular area despite greater real social costs, causing some of the locational costs of residing in the area to be borne by others outside the area. Warford's reluctant conclusion seems to be that when there already exists a constellation of subsidies across other services, it hardly seems possible to have their consumers bear the full burden of water and sewerage costs without subsidy. Since we have already twisted the efficient distribution of population among places within the country, how can we ask those so trapped by their location to pay the full cost of a clean water supply?

This seems to suggest that there may be times when a subsidy will occur even though the activity is unwarranted from the perspective of benefits and costs. Taking the case of state intervention in situations of bankruptcy, it could be argued on the grounds of efficiency that intervention is unwarranted unless it can be proven that benefits are greater than costs. On the other hand, the request for intervention is made primarily because revenues fall short of expenditures.

In this case, if water system externalities are internalized, for example, via EPA and state regulations requiring the clean-up of effluent, then the firm's private cost function closely approximates the true social cost function. Since consumer prices presumably cut off demand at points where the marginal benefits of the last units of water equal the price, there is an apparent paradox. Consider figure 3–1.

Here $ac > ar$ at the point of maximum profit (more to the immediate point, minimum loss: the point where $MR = LRMC$). This is the rationale for claiming bankruptcy, but if in some sense the benefits of this are not greater than the costs, then the output should not be produced. The system should be disbanded and no subsidy given.

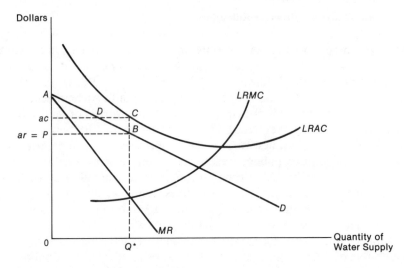

$LRMC$ = long-run marginal cost; $LRAC$ = long-run average cost; D = demand; MR = marginal revenue.

Figure 3-1. Potential Water Supply Cost and Demand Conditions

Upon consideration, recognizing that the benefits of consumption (what people would have been willing to pay) are more than what they have paid, these benefits are measured by the area under the demand curve. At output Q^* the benefits then are area $OABQ^*$. Since costs are area $O \cdot ac \cdot C \cdot Q^*$, then if area $A \cdot D \cdot ac$ > area BCD, benefits exceed costs, and the subsidy is warranted.

It is useful to point out, however, that this still does not imply that a subsidy is necessary. If the firm could get everyone to pay the full value for all the water that they consumed, the differential sets of prices would represent economic price discrimination. But with this discrimination, the firm can raise average revenues to or above costs. Effectively, in the limit the demand curve becomes the marginal revenue curve, and the average revenue curve rises above it as shown in figure 3-2. Under such conditions the social optimum occurs where price equals marginal revenue at Q^{**}, and at that point average revenue is above average cost.

Whether chronic loss making is resolved either by giving a firm subsidies or by allowing a firm to increase the discrimination in its pricing, it is clear that there will be interpersonal equity transfers. It is hard to classify these distributional shifts uniquely, however, because they depend on which solution is chosen. If discriminatory pricing obtains, then the distributional shifts are determined be the relative trend of personal demands for water. If subsidies obtain, then the distributional shifts depend on the nature of the subsidy and how it is used by the firm.

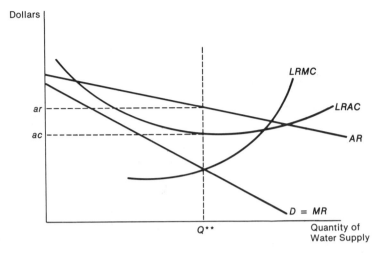

Figure 3–2. Alternative Pricing Conditions in Water Supply

Distributional Impacts of Service Extensions
from Subsidy Programs

Much of the impact of clean water subsidies is derived from the system expansions that accompany investment. In wastewater programs the extension of service followed directly the five-fourths' rule, and the fact that only interceptor sewers, not laterals, could be funded. That communities used these funds as a source of development capital implied that their use was greatly directed toward the extension of wastewater service through undeveloped urban fringe areas. In water supply treatment programs, service extension has followed the twenty years' growth rule and the fact that the necessary capital construction costs would be reduced on average if the base of consuming population was increased. While the federal funds cannot be used directly for developing the extensions themselves (until the 1981 amendments to the Safe Drinking Water Act become effective in 1984), they can and will be used in increasing treatment capacity to accommodate such extensions.

The distributional impacts of such extensions are a little easier to trace. The distributional effects are embedded primarily in the housing market because water system extension ties or binds water service to specific properties. Some of the benefits of the newly acquired service are retained by the property owner since he or she is also a consumer of water service (abstracting from rental markets). Other benefits that are trapped in increasing property prices can be thought of as being obtained by the property developer.

Consider a community that adds a source of clean water to properties in

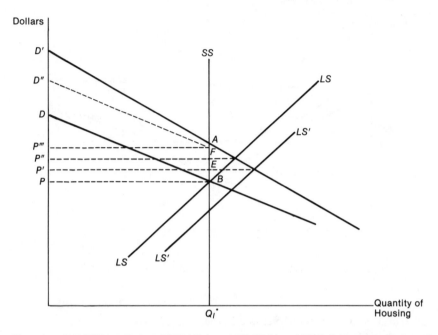

Note: Area $D''ABD$ equals area $ABPP'''$ since $\Delta D''AP''' \approx \Delta DBP$. Subtracting the area of $\Delta DCP'''$ and adding the area of ΔABC to the areas of $\Delta D''AP'''$ and ΔDBP, respectively, proves the statement.

Figure 3–3. The Property Market for Housing which Receives Water
              Investments

place I. As figure 3–3 suggests, the demand for such property shifts upward (from D to D'). This happens in proportion to the extent to which the water market does not capture everyone's total willingness to pay for water service in user charges. Specifically, any consumers' surpluses are transferred as a present value of the expected consumer surplus annuities.

For a market reflecting the numbers of residential units borne by the fixed land area in parcel I, long-run supply will be upward sloping with some elasticity if in fact suppliers can subdivide, for example, along a long-run supply function LS. If no subdivision is possible or if we are looking at the short run, then the supply will be inelastic along short-run supply SS. If the water extension makes it easier to subdivide because of not needing on-site supply or disposal, then the supply curve shifts to a new supply curve LS. In any case the price of housing is likely to shift upward to either P', P'', or P''', depending on supply conditions.

Looking only at the initially existing Q_I^* properties, it is now possible to determine the distributional shifts of such benefits. Drawing D'' parallel to

D, then the area $D'AD''$ represents a benefit of the investment in clean water that accrues to the owner-resident. This is not transferred in the property market because it is not reflected in increasing property values even when these capture the greatest effect under inelastic short-run supply conditions. Beyond the $D'AD''$, the balance of the clean water investment benefit is $D''ABD$. It is transferred entirely and precisely to the property and absorbed by the developer-speculator if supply condition SS obtains. It is transferred less than entirely if LS or LS' obtains. In these cases $BFP''P$ or $BEP'P$ are transferred respectively to the property market developer. The respective balances, $AFP''P'''$ or $AEP'P'''$, are retained by the owner-consumer in addition to area $D'AD''$, already described to be retained by him or her.

Since windfall profits in the housing market accrue to the developer whenever water or sewerage is extended, it is clear that the developer will push for such investment. Further, since such developmental profits depend on both the number and the sizes of the properties that are so developed, it is likely that the developer may contrive to influence the kind of development that takes place. The motive is sharper the higher the return to the developer.

Empirical Measurement of Distributional Impacts

There are certain conditions under which it is relatively easy to determine the distributional impacts of investment in clean water. These occur when there are few other confounding macromarket effects that simultaneously influence property prices other than the investment itself.

Under the conditions in this simple case, it is possible to run a cross-sectional, or "hedonic," index of the impact of availability of supply on property value. This holds constant lot size and other variables, identifying the value of the clean water investment in the coefficient of a dummy variable reflecting the change in water and/or sewerage supply from on- to off-site. This effectively measures the difference $P''' - P$. It is similarly possible to run a time-series index of the impact of the availability of water supply extension on property values. Failing to hold constant lot size and other variables that are influenced by the clean water investment/extension into new properties, this technique identifies the equilibrium values that actually obtain, P''', P'' or P', depending on supply conditions.

Between the two techniques, it is possible to identify what the property value changes would necessarily have been if it were to have measured the fully transferred value of the benefits of clean water investment and what actually occurs. The difference between these amounts measures the return to the owner-consumer of property and clean water service. The increase in the times-series measures the return to the developer-speculator. Comparing the two gives one the distribution of benefits to the consumer and the developer.

There is a significant caveat in all of this analysis, however. Given the conventional measurement of time-series property value indices, it is unlikely that the results can be interpreted simply. Namely, the more there are other macromarket property effects influencing the property market time series, the more one will need to extract these influences with the use of techniques such as those developed by Nourse;[5] Ridker and Henning;[6] and others who use the Bailey, Muth, and Nourse methodology[7] for time-series analysis of property markets.

Consider a region with three subsectors, I, J, and K. Sector I is the project area, recently receiving water and sewerage service. Sector J has never had the service, and within relevant history, K has always been serviced. Time 1, noted as a subscript, is before the project is implemented in I. Time 2 is after the project. We assume periods 1 and 2 are sufficiently removed from the project that they represent the last and first periods of general equilibrium before and after the project, respectively. Hence they are sufficiently removed that they are devoid of anticipations and partial adjustments.

Properties I, J, and K are assumed to be highly substitutable, with the exception of their water and sewerage, and thus they are taken to be in different markets. Initially, I is indistinguishable from J, and it later becomes indistinguishable from K so the system of properties is divided into two markets, not three. While we assume away the possibility of rezoning that often accompanies water and sewerage investment, that may be treated explicitly (and is treated in Coelen and Carroll[8]). We posit that as I properties leave the J market, they maintain their physical, numerical quantity in units. They do, however, obtain a greater facility for subdivision, and that is reflected in an increased elasticity to their long-run supply.

Assume reasonable competition so that conditions dictate the equivalence of prices within markets at a given time. The prices of I and J are equalized in the first period, as are I and K prices in the second period:

$$_IP_1 = {_JP_1} \quad \text{and} \quad {_IP_2} = {_KP_2}$$

At any time, as Rosen has shown,[9] each of these prices indicate the value of the full range of hedonic characteristics associated with the property types.

Believing that the cross-sectional differences in prices between housing that differs with respect to a single characteristic can be identified as the hedonic value H of that characteristic, there are clearly two distinct hedonic values. One is calculated before and one after the project:

$$H_1 = {}_KP_1 - {}_JP_1 \equiv {}_KP_1 - {}_IP_1$$

and

$$H_2 = {}_KP_2 - {}_JP_2 \equiv {}_IP_2 - {}_JP_2$$

These may be identical but need not be. Ambiguity is inherent between them because of the possible shifts in ${}_KP_t$ and ${}_JP_t$ over time. Specifically, the supply adjustments in these markets may imply that ${}_KP_1 \neq {}_KP_2$ and ${}_JP_1 \neq {}_JP_2$.

If a time-series, control area technique is used on the same data, then the time-series indicator of value shift TS is also ambiguous because either serviced or nonserviced properties may be taken as the control. We can define the possibilities:

$$_nTS = {}_IP_2 - {}_IP_1 - ({}_JP_2 - {}_JP_1)$$

and

$$_sTS = {}_IP_2 - {}_IP_1 - ({}_KP_2 - {}_KP_1).$$

Using conditions (1), appropriate prices are substituted for the ${}_IP_t$,

$$_nTS = {}_KP_2 - {}_JP_1 - {}_JP_2 + {}_JP_1 = {}_KP_2 - {}_JP_2 = H$$

$$_sTS = {}_KP_2 - {}_JP_1 - {}_KP_2 + {}_KP_1 = {}_KP_1 - {}_JP_1 = H_1$$

which holds, unequivocally, regardless of what kinds of property value reactions there may be to supply changes in J and K. It can be shown, however, that $_nTS$ and $_sTS$ differ from the true time-series change in property, without control effects $TS = ({}_IP_2 - {}_IP_1)$, depending on the extent of offsetting J and K prices:

$$_nTS = TS \text{ iff } {}_JP_1 = {}_JP_2$$

and

$$_sTS = TS \text{ iff } {}_KP_1 = {}_KP_2$$

Consequently, a time-series method will estimate the actual value change if and only if we choose the control area with sufficient size that it may absorb the supply impact of shifting I without price adjustment. This concurs with the findings of Polinsky and Shavell.[10] Further, and most important for the issue at hand, time-series methods are identical to the cross-section as long as a correspondence is appropriately qualified between control area (serviced/nonserviced) and pre- or post-hedonic value.

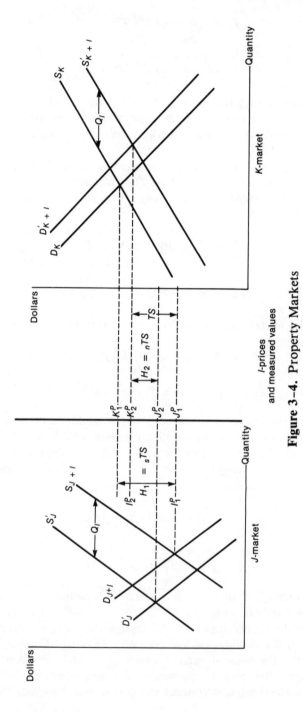

Figure 3–4. Property Markets

Figure 3–4 shows graphically, for the most general case, the shifting of values in both J and K. The quantity Q_1 is the number of I properties shifting from J to K. S'_J and $S_{J + I}$, supplies in the J market, are more inelastic than S_K and $S'_{K + I}$, supplies in the K market, because of easier subdivision. Transfer of I from J to K has no effect on remaining ease of subdivision (accounting for the parallel shifts in supplies). Since housing is tied to the property on which water supply is provided demands shift in response to the project. Yet, many residents retain an interest in I-area properties even as these properties shift from the J to K markets because these residents may have large existing consumer surpluses. In addition, there may be "information" effects or changes in preferences resulting from a project that also account for demand shifts. All these issues must ultimately be resolved before the question of degree of capitalization can be answered. Our point here, however, remains that our time-series information yields conceptually identical results to the cross-section regardless of capitalization, under several specific forms that it may be implemented.

Empirical Results and Policy Conclusions

The best results that we have obtained so far is that an apparent 25 percent or so of the full value of water supply benefits is trapped in user charges. This result obtains for developing country cases, as demonstrated in work done for the World Bank by Bahl and Coelen, in the early 1970s.[11] Of the remaining 75 percent, approximately another 75 percent of it may be captured by the developer with the balance going to the consumer. This result, obtained for a domestic case study for the Army Corps of Engineers, was noted in a final report by Aron and Coelen.[12] Altogether then, 50–55 percent goes to the developer, and 25 percent goes to the utility in paying costs of providing the supply. The remaining portion of approximately 20 percent remains with the consumer as a true consumer's surplus.

The result of this distribution suggests that there are heavy developmental pressures that come to bear on the process of funding clean water. The more these pressures can be eliminated, the more direct and appropriate the program for funding clean water becomes. Placing a developmental tax on the increased property value, as New Jersey did,[13] to capture windfall property returns, limits the kind of growth that will be tolerated by water system extension. In a different way, the 1981 amendments to the Safe Drinking Water Act does the same. Stricter enforcement of the distributional accounting requirement in the 1972 Water Resource Council standards and guidelines would help identify the problem on a case-by-case basis. Still other, new measures, however, will surely be required to help rationalize the process.

Notes

1. C. Binkley, et al. *Interceptor Sewers and Urban Sprawl* (Lexington, Mass.: D.C. Heath and Co., 1975).

2. M. Morgan, "Guide to State and Federal Policies and Practices in Rural Water-Sewer Development," Commission on Rural Water, 1974.

3. H. Averch and L. Johnson, "Behavior of the Firm under Regulatory Constraint," *American Economic Review* 52 (December 1962): 1053–1069.

4. J.J. Warford, *The South Atcham Scheme: An Economic Appraisal,* Her Majesty's Stationary Office, London, 1969.

5. H.O. Nourse, *The Effect of Public Housing on Property Values in St. Louis* (Ph.D. diss., Department of Economics, University of Chicago, August 1962).

6. R.G. Ridker and J.A. Henning, "The Determinants of Residential Property Values with Special Reference to Air Pollution," *Review of Economics and Statistics* 69 (no. 2) (May 1967):246–257.

7. M.J. Bailey, R.F. Muth, and H.O. Nourse, "A Regression Method for Real Estate Price Index Construction," *Journal of the American Statistical Association* 58 (December 1963):933–942.

8. S.P. Coelen and W.J. Carroll, *Residential Location and Urban Housing Markets: Studies in Income and Wealth* 43, edited by Gregory Ingram, NBER, 1977, pp. 381–389.

9. S. Rosen, "Hedonic Prices and Implicit Markets: Production Differentials in Pure Competition," *Journal of Political Economics* 82 (1974): 34–55.

10. A.M. Polinsky and S. Shavell, "Amenities and Property Values in a General Equilibrium Model of an Urban Area," *Journal of Public Economics* 5 (1976):119–130.

11. R. Bahl and S. Coelen, "Water/Sewage Investments and Land Values-Nairobi and Kuala Lumpur Experience," final report to the International Bank for Reconstruction and Development, 1973.

12. G. Aron and S.P. Coelen, "Economic and Technical Considerations of Regional Water Supply," final report to the Institute of Water Resources, U.S. Army Corps of Engineers, Contract Report 77-7, July 1977.

13. New Jersey Department of Community Affairs, *Secondary Impact of Regional Sewerage Systems,* Vol. 1, June 1975.

4

The Federal Role in Funding Clean Water

J. Leonard Ledbetter

The continuing debate regarding the appropriate role and level of funding by the federal government under the Clean Water Act surfaced before enactment of the legislation in 1972. Reflecting the Nixon administration's opposition to the act, John Quarles, deputy administrator of the U.S. Environmental Protection Agency (EPA), stated at the time that "state agencies appeared hopelessly inadequate in terms of their manpower, equipment, and expertise to meet the challenge of effectively controlling complicated pollution problems. To do the job, it was felt, one had to turn to the federal government." Observing the progress made during the ten-year period 1972–1981, however, Quarles noted one major difference: "In particular, the capacity of state programs to carry major responsibilities in administering ambitious regulatory requirements is greatly enhanced." He further argued that the field of pollution control was dominated by an unneeded federal intrusiveness.[1]

Congress has continued to express the intent to maintain a federal role in funding clean water. This role was defined in the Clean Water Act in Section 101(b), which notes, "It is further the policy of Congress to support and aid research relating to the prevention, reduction, and elimination of pollution and to provide Federal technical services and financial aid to State and interstate agencies and municipalities in connection with the prevention, reduction, and elimination of pollution."[2] In 1980, following an extensive examination of the wastewater treatment facilities construction program, the Subcommittee on Investigations and Oversight of the Committee on Public Works and Transportation of the U.S. House of Representatives restated the need for continuation of federal funding, concluding that "the construction grants program should be continued for at least 10 years, and that there is a continuing federal interest arising from the interstate nature of the waters affected and the implications for public health associated with pollution."[3]

Authorization versus Appropriation

Historically, however, Congress has been more inclined to authorize funding for the clean water program than to actually appropriate such funds—a situation that preceded the enactment of the Clean Water Act. Federal budgetary commitments in other areas such as to the Vietnam war, resulted in less than full funding of clean water programs from the mid-1960s to 1971.

Presidential policy on funding clean water has varied considerably. The Nixon administration attempted unsuccessfully to veto the Clean Water Act of 1972 and also to reduce and/or delay funding for the program. President Ford announced that he would support funding of a construction grant program of $45 billion over a ten-year period. President Carter indicated his support for an equivalent federal outlay over a ten-year period; however, as he proceeded with efforts to balance the federal budget, he decided to reduce the appropriation level for the time he was in office.

For these and other reasons, considerable instability and fluctuation have occurred in the appropriated level of funds since FY 1973, as can be seen in the following table:

Construction Grant Funds
(billions of dollars)

Fiscal Year	Authorization	Appropriation
1973	5.0	2.0
1974	6.0	3.0
1975	7.0	4.0
1976	0.0	9.0
1977	1.48	1.48
1978	4.5	4.5
1979	5.0	4.2
1980	5.0	3.4
1981	5.0	2.55
1982	2.4	2.4
1983	2.4	2.4

In 1982 the Water Pollution Control Federation (WPCF) developed and published the position paper "Funding National Water Pollution Control Programs." In its recommendations, the paper stated: "Congress should consider other ways to fulfill the national interests of the Clean Water Act, including trust funds, block grants, and low interest loan programs. Congress must set an endpoint to the construction grants program. It should then provide a reasonable and predictable level of funding to the program until that endpoint is reached.[4] This position of supporting a reasonable but limited federal involvement in the construction of needed municipal wastewater treatment facilities was encouraged by other groups as well.

During 1981–1982, the states coordinated their positions on the issue of federal funding under the Clean Water Act through the Association of State and Interstate Water Pollution Control Administrators (ASIWPCA), the

National Governors' Association (NGA), the National League of Cities (NLC), and other groups to impress upon the Congress that the federal mandates in the Clean Water Act necessitated a federal budgetary exposure. The 1981 changes to the Clean Water Act, entitled "Municipal Wastewater Treatment Construction Grant Amendment of 1981," reflected the involvement of all these groups and many others. It was clearly understood in the lengthy discussions and compromise on the legislation that the Reagan administration was accepting the concept of a federal commitment of approximately $32–36 billion over at least a ten-year period. As a part of the compromise, the state and local governments accepted a lower annual authorization for funds for construction grants. Equally significant, at the end of FY 1984 the federal share decreases from 75 percent to 55 percent. Section 10 of the 1981 amendments, entitled "Reserve Capacity," states that no grant after October 1, 1984, shall be made to construct that portion of any treatment works providing reserve capacity in excess of existing needs on the date of approval of a grant for construction.[5]

New Federalism

Beginning in early 1982 the Reagan administration placed great emphasis on its "New Federalism" and announced specific programs scheduled for inclusion. The construction grant program for wastewater treatment facilities is now being proposed by the administration as one of the programs to be included for funding out of the trust fund outlined under the New Federalism strategy. This concept has created considerable confusion as to the future of federal funding for construction grants.[6]

The administration has proposed to reduce federal support of the clean water program in other areas as well. A proposal to phase out the federal annual financial assistance to the states for the management of the state water pollution control program was outlined. This program assistance is authorized and funded at approximately $54 million per year, and a reduction or elimination would be severely detrimental to the nation's water quality program. Significant increases through state funding have been provided over the past ten years for these programs for the overall administration of the water quality/water pollution control efforts; however, the inflationary costs during that period greatly eroded the impact of the increased state appropriations.

Reductions in the EPA budget for research programs and administrative programs have been proposed in recent years, and Congress reduced the level of funding for those programs for FY 1983. Recent funding levels are in the table that follows:

EPA Budget

(millions of dollars)

	FY 1981	FY 1982	FY 1983
State management grants (Section 106)	51.2	51.2	54.1
Clean lakes	11.0	9.0	3.0
EPA			
Salaries and expenses	562.0	555.0	548.6
Research and development	251.0	154.3	119.0
Abatement and control	534.8	373.0	369.0

Responsibilities and Funding

The assignment of responsibilities under the Clean Water Act has involved numerous groups, including all levels of government. Immediately following enactment of the act in 1972, Congress transferred responsibilities from the states to the federal government by preempting the states in areas such as permit issuance for point sources, non-point source pollution control, and construction grant management. Extensive preemption and transfer of responsibilities from the states resulted in confusion, delays in implementation of the objectives of the legislation, and negative overall results.[7]

In determining the source of funding for the administration of an activity such as the clean water program, an initial effort should be made to identify the level of government at which a specific function can be most efficiently administered. An evaluation should then be conducted to establish the costs mandated by the implementation of the various functions in order to allocate the funding appropriately.

Recognition of the efficiency of several functions being conducted at the national level, along with the essential need for equitable national standards applicable to all benefactors of the water resources, should result in a logical assignment of responsibilities. Federal government agencies should oversee (1) research, (2) developments of effluent and ambient standards, (3) ocean-dumping control, (4) resolution of interstate issues, (5) overview of state programs, and (6) technical assistance.

States should be responsible for (1) waste load allocation, (2) issuance of permits, (3) water quality management, (4) compliance and enforcement, (5) ambient monitoring, (6) construction grants management, (7) operator training, and (8) emergency response activities.

Finally, local government should oversee (1) wastewater facility design, (2) construction of facilities, (3) operation and maintenance, (4) establishment of user fees, (5) pretreatment requirements, and (6) effluent monitoring and reporting.

The assignment of responsibilities and the allocation of costs should address several factors, namely:

1. An evaluation and identification of the responsibilities to be assigned should be conducted.
2. The level of government at which an activity can be most efficiently administered should be determined.
3. Duplication of effort or encroachment by one agency into another's responsibility should be prevented.
4. Responsibilities should be identified and assigned with as much flexibility as possible to accommodate the capabilities of each level of government.
5. Performance audits are needed to confirm whether assigned responsibilities are being funded and conducted.

The identification of roles and the assignment of responsibilities will not by themselves resolve the issues of funding, but the preceding procedure could assist in arriving at an equitable and realistic funding strategy. For example, it will be more cost effective for the national government to fund and direct research programs. This should not preclude the federal government from using the private sector and educational institutions to conduct the research programs, but the prioritization and overall direction of the research efforts on a national level eliminate unnecessary duplication.

Funding on the Wane?

Numerous national and local polls have been conducted regarding the interest and support of the American public for environmental programs. These polls consistently reveal a strong support for clean air and clean water programs. During 1982, Congressman Elliott Levitas of Georgia, chairman of the Subcommittee on Investigations and Oversight of the Committee on Public Works and Transportation, conducted a public opinion poll on the subject. Over 70 percent of those responding indicated support for a strong national clean water program and a willingness to pay more in order to assure continuation of the program. Similar results to the Levitas poll have been reported from several areas of the country. Nevertheless, there has been a decrease, through inflationary impacts and reduced appropriations, for programs funded under the federal Clean Water Act.

Why has this occurred? Why, contrary to the public's position, does there appear to be less support and enthusiasm on the part of Congress and the administration for funding the clean water program? Some frequent responses include:

Unattainable goals were established.

The clean water program suffers from a negative public works program image.

The clean water program was adversely impacted with an overkill of regulations.

Progress has been difficult to identify and quantify.

A complex problem was oversimplified.

There has been inconsistent enforcement of standards.

To make all waters fishable and swimmable and attain "zero discharge of pollutants" may be a desirable goal, but in my opinion with the available technology and funding, it must be recognized that it is unattainable. Although it may be possible to accomplish a level of water quality that would allow movement of fish through all major water bodies and reproduction in most of them, is it necessary to attain that quality of water in small streams below small communities where the stream is in fact intermittent—or would be without the treated effluent from the community? A degree of realism is greatly needed in the program.

The large sum of federal funds authorized and partially appropriated has created great expectations, and the gap between authorization and appropriation levels has spawned problems. Realistic funding and scheduling has been extremely difficult to achieve. In addition, the magnitude of the federally funded program attracted many newcomers to the water pollution control field. The large number of persons, firms, equipment suppliers, and so on moving in and out of the water pollution control field is partially reflected by reviewing the recent report of the Water Pollution Control Federation (WPCF) Long Range Planning Committee. In that report, "Evaluation of WPCF Membership, Recent Past and Near-Term Future," the committee reported significant turnover in membership of WPCF with a trend of no growth. While many expected the authorized funding to create an accelerated public works program, the actual funding, coupled with the regulatory influence, has dampened the enthusiasm of many members of Congress, state and local officials, and others.

The dictates of Congress, combined with other actions on the national scene, has resulted in an overkill of regulatory requirements being imposed under the Clean Water Act. Effluent requirements, levels of wastewater treatment, water quality standards, construction grant procedures, and unproductive, expensive planning efforts have contributed to confusion and loss of support for the clean water program. In some cases, unrealistic levels of wastewater treatment have been required, while in others, excessive delays have been fostered by the regulatory process. The maze of regulations was enough to stifle the program, but in addition to the quantity of regulations, there has been continuing change in the regulations.

Success stories related to the Clean Water Act of 1972 are difficult to identify and quantify. In Georgia much of the work was under construction or already in place to correct water pollution problems prior to the Clean Water Act. All the industries had water pollution control facilities installed or under construction before the Clean Water Act was passed. Very few of these industries needed to make significant modifications in order to comply with NPDES requirements. Since Georgia is the leading state in pulp and paper production, the major producer of tufted carpets, one of the leaders in textiles, as well as food processing, industrial wastewater treatment is a significant element of the state's water quality program. However, the progess in cleaning up the state's waters is not attributable to the Clean Water Act—that progress was made prior to the Clean Water Act.

Most of the progress and anticipated success stories that can be attributed to the Clean Water Act are projects that are under construction. When the projects under construction in the Atlanta metropolitan area are completed, successful and significant improvement in water quality will be reflected. In other communities where wastewater systems serve both industry and the local population, definite progress will be measureable upon completion of the projects that have been partially funded by the Clean Water Act.

Ironically, the confusion related to older publicly owned treatment works (POTWs) not meeting the National Pollution Discharge Elimination System (NPDES) requirements has reflected negatively on the clean water program. For example, Atlanta's large POTW was not funded under the Clean Water Act nor designed to meet the requirements of the EPA rules for secondary treatment facilities. The Atlanta POTW was designed prior to the Clean Water Act and was not constructed to meet the permit limits that were imposed by the NPDES permit program. The attention that was focused on this plant for failure to meet the permit as well as the inadequate level of operation and maintenance by the city contributed to negative attitudes toward the clean water program by some members of Congress as well as others. It is unfortunate that the pre-Clean Water Act and post-Clean Water Act periods have been confused by many, including the Government Accounting Office.

The complexity of the clean water program has been submerged in many instances by various groups in their efforts to oversimplify the program. An overindulgence in planning in the early days of the clean water program, too much emphasis on large regional systems, contention that non-point sources were greater problems than point sources, and debates on the issue of water quality standards or effluent limitations are handy examples of such simplification. Attention to issues such as combined sewer overflows, infiltration/inflow, innovative and alternative technology, advanced wastewater treatment, and cost effectiveness contributes to the complexity of the clean water program. However, the inclination to isolate these issues and address them independently has added to the dismay and frustration associated with the program.

Inconsistent enforcement by EPA and by many of the states has been a factor in the loss of enthusiasm for the clean water program. To cite one Georgia example, a company manufacturing titanium dioxide using the sulfate process was required to install a modern water pollution control facility in the mid-1970s. The other five or six similar plants located in other states have not yet installed adequate, modern water pollution control facilities—yet the Georgia manufacturer must compete with them. Consequently, the Georgia company continues to report an annual financial loss, a major cause being the expenses incurred related to wastewater treatment.

For the most part, there has been considerable pressure on industry to install water pollution control facilities. That has not been the case for local governments. Congress even added language to Section 301 of the Clean Water Act in 1977 to allow existing POTWs to delay meeting requirements of the act, "based on the earliest date by which financial assistance will be available from the Federal Government." In reality, the Clean Water Act established a double standard for pollution abatement—one for industry and the other for local governments. If an industry is served by the POTW in the community, the double standard is even more complicated. This approach to enforcement of the national program has contributed further to loss of enthusiasm for the clean water program.

As a result of the current situation, it should be no surprise to hear that the direction of the program for this decade is unclear. The factors that may "drive" the program or hinder it are many. Some of the more pertinent factors that can be expected to surface as significant are:

1. *Funding*
   a. Publicly owned treatment works (POTWs)
   b. The "user pays" concept

2 *Emphasis on improved management*
   a. Financial resources
   b. Existing wastewater systems
   c. Greater accountability

3. *Allocation of water resources*
   a. States competing on interstate streams
   b. Communities competing on a stream
   c. Various types of water users competing
   d. Future economic development

4. *Reassessment of water quality standards and effluent requirements*
   a. Little support for "clean up at any cost"
   b. Relationship of cost effectiveness to goals of Clean Water Act
   c. More flexibility to states to evaluate downstream uses and needs
   d. Resolution of the double standard applied to industries and POTWs

5. *Water quality, water quantity, surface waters, and groundwater need coordinated attention*
   a. Recognize interdependence
   b. Protect pathway to man

The preceding factors, the emphasis on reductions in federal spending, and the overall slowdown in the nation's economy have definitely caused a current waning in the funding for clean water programs. The combination of these factors make it essential for the professionals and others interested in the field of water pollution control and water resource management to clarify the specific roles of each level of government and assure appropriate corresponding levels of funding.

Affordability versus Accountability

Considerable concern has been expressed throughout the duration of the clean water program regarding the concept of a massive construction grants program rewarding the recalcitrant. In addition, the grant program has not effectively distinguished the difference between cost and affordability. A community has been held accountable for the installation of whatever amount (capacity) and degree of treatment (secondary or advanced treatment) that was deemed necessary by their consultant, the state, or the EPA. With a 75 percent federal grant available, most local officials have not objected to being held accountable for very high levels of treatment. In several cases it has now been established that some communities, especially smaller ones, do not have the financial resources to afford the required treatment works because of the high costs of operation and maintenance. Yet a costly project for a large population center may cost each person or household a very acceptable level of increase in their utility bill. In addressing this complex issue, the report by the Subcommittee on Investigations and Oversight of the Committee on Public Works and Transportation contained the following excerpt:

> For example, in Atlanta, it will cost $50 million to upgrade the City's R.M. Clayton treatment plant pursuant to State and EPA requirements. Yet, after the EPA grant, this will cost only 67 cents per household per month.

Added to a $10 per month current cost, this will be very "affordable" to most, if not all, area residents.

In contrast, a less costly project for a small community may be virtually unaffordable. Consider, for example, the plight of a small New Hampshire community whose 750 residents face a $24 a month charge for their sewer service, despite Federal and State grants which covered 95 percent of their construction cost.

In part, this disparity results from the economy of scale present in a large system, and the lack of it in a small community. However, there are other factors that affect these costs, most notably the environmental sensitivity of a community's location and the resulting degree of treatment that may be required.

Until such time as this cost-burden "inequality" characteristic of the program is addressed, residents of small communities will have to pay a greater share of their personal income for cleanup costs than those in larger areas.

The combination of the reduction in federal funding and the scope of a project eligible for funding assistance necessitates further evaluation of the affordability issue, as well as alternative financing strategies. Since states are absorbing significant percentages of several previously funded or partially funded federal programs, limited-to-insignificant state funding for water pollution control projects can be expected in most cases. Therefore, it is timely to examine the alternatives available for a shift from federal funding as well as the affordability of required treatment levels.

Two examples will be reviewed briefly. On the subject of affordability of the user where a 75 percent federal grant was available, a comparison can be made for the costs to the customer in relation to other utilities. The following table reflects that reasonable costs, compared to other utilities, exist for a community of approximately 500,000 population.

De-Kalb County, Georgia
Wastewater Treatment Costs versus Utility Costs

Utility/Service	Firm	*Expenditures per Household ($/Year)*
Electricity	Georgia Power Co.	438
Gas	Atlanta Gas Light Co.	380
Telephone	Southern Bell	306
Garbage collection	DeKalb County	97
Cable TV (includes one movie channel)	Cable DeKalb	228
Water	DeKalb County	45
Wastewater (after implementation of modern wastewater facility)	DeKalb County	90

While the preceding tabulation demonstrates a reasonable annual cost for one community, it must be emphasized that a 75 percent federal grant was used to construct the project, which had a total cost of about $60 million. Therefore, the other example to be considered relates to the importance of the establishment by the states of some type of alternative financing where only insignificant federal funding (or none at all) will be available. In this case the state of Georgia enacted legislation in 1982 creating the Environmental Facilities Study Commission, charging it with the responsibility of examining the subject of alternative financing and the development of a report.[8]

The Georgia study revealed that for local governments to meet current environmental requirements would require, on the average, that water/sewer rates be doubled or tripled. These predictions were based on the most optimistic assumptions of federal fund availability and the elimination of inflation and rapid population growth factors. Therefore, for most communities, the water/sewer rates will increase severalfold without assistance from the federal and state governments. The affordability versus accountability aspect of the clean water program can best be addressed by a continuing commitment by all three levels of government.

Federal Role Elements

The federal role in funding clean water is usually considered in terms of direct grants to the states, construction grants to local governments, and EPA administrative costs. Yet there are other significant elements of the federal role. Throughout the nation there are large tracts of federal lands that must be managed in a manner to minimize any degradation of water quality, and in many cases these lands should be managed to enhance water quality. Proper management of these areas must be recognized as an essential and continuing federal cost.

In addition, numerous federal facilities, such as military installations and large dams, are scattered throughout the United States. Too frequently in the past these facilities have been slow to comply with the requirements of the clean water program. Again, a continuing federal commitment for the appropriate level of funding for such projects remains essential to the clean water program.

The federal role must also include an element to better inform and educate the citizens on proper utilization and conservation of the nation's valuable water resources. National efforts on education can be more cost-effective and should be expanded as one of the key elements of the federal role in the future.

Another element of the federal role that impacts funding is the establishment of treatment levels and water quality standards. Through the authority provided by Congress under the Clean Water Act, the EPA determines the effluent limitations and participates in the development and establishment of state water quality standards. Since these two factors in turn impact the cost of wastewater treatment by industry and local governments, greater flexibility should be provided by EPA to the states to assure the installation of cost-effective systems while still protecting water quality. This element of the federal role in relation to costs or funding has not been utilized appropriately in the past ten years to minimize costs and expedite pollution abatement.

Conclusions

1.   There is a continuing federal interest arising from the interstate nature of the affected waters and the implications for public health associated with water pollution. This should be recognized, and the federal funding provided to the states and local governments should be commensurate with these interests.

2.   There is a need to evaluate and clarify the financial impact on the state and local governments of federal mandates under the Clean Water Act. Then federal funds should be provided on a continuing basis to help fund those mandates.

3.   Research and development, technical assistance, educational efforts, delegation and oversight, general administrative costs, and related activities necessitate a continuing federal involvement in the clean water program, as well as the funding to assure effective results.

4.   The federal funding role needs to be stabilized to assure the most cost-effective impact on the clean water program for the future. Fluctuations and proposed changes in funding levels have been detrimental to the clean water efforts over the past ten years.

5.   Administration of the various elements of the clean water program at the most efficient level of government should identify an equitable and realistic funding strategy, including the federal role.

6.   A reduction in funding of the clean water program at the federal level can be attributed to several factors, including the establishment of unattainable goals, overregulation, inconsistent enforcement, and lack of progress identification. A national educational effort should be implemented to assure better understanding of the issues and the continuation of an appropriate federal funding level.

7.   The affordability versus accountability aspect of the clean water program can best be addressed by a continuing commitment by all levels of government.

8.   The federal role in funding clean water includes: direct support to the states and local governments; management of federal lands and facilities; public education; research and administrative programs; and, especially, the establishment of realistic and cost-effective requirements.

Notes

1. The Academy of Natural Sciences, "Environmental Control and the Role of the Federal Government," *Proceedings of the National Water Conference,* ed. by James Wilson, Philadephia, Pa., 1982, p. 153.

2. *The Clean Water Act,* as amended through 1981.

3. House Subcommittee on Investigations and Oversight of the Committee on Public Works and Transportation, *Implementation of the Federal Water Pollution Control Act Concerning the Performance of the Municipal Wastewater Treatment Construction Grants Program,* Washington, D.C., 1981.

4. Water Pollution Control Federation, "Funding National Water Pollution Control Programs," *Journal of the Water Pollution Control Federation,* April 1981, p. 433.

5. *Municipal Wastewater Treatment Construction Grant Amendments of 1981.*

6. Municipal Finance Officers Association, *Financing Water Pollution Control: The State Role,* Chicago, August 1982.

7. The Academy of Natural Sciences, "How Do You Transfer Responsibilities?" *Proceedings of the National Water Conference,* Philadelphia, Pa., 1982.

8. Georgia Environmental Facilities Study Commission, *Options for State Financial Assistance to Local Governments for Environmental Facilities,* September 1982.

5 State and Local Roles in Funding Clean Water

Robbi J. Savage

At their 1980 annual meeting, members of the Association of State and Interstate Water Pollution Control Administrators (ASIWPCA) heard Governor Richard Snelling of Vermont warn: "The buck has stopped on your desk and it's a smaller buck, a more tightly controlled buck, than the authors of the 1972 [Clean Water Act] Amendments or the environmental visionaries of the seventies ever dreamed of."

Governor Snelling went on to emphasize that state water pollution control programs would have to operate with fewer federal dollars in the future because of increasing pressures to control the size and cost of the federal government. He further contended that states could manage effectively at these lower funding levels by operating less complicated programs that avoided the wasteful paperwork and excessive regulation that too often hampers federal programs.

These remarks demonstrate that the states have realized for several years—well before the election of the current administration—that the federal role in funding clean water was on the decline and that the state and local role would have to expand accordingly. This administration's budget cutbacks and New Federalism policy directions are simply continuations of trends already emerging in the late 1970s.

Clean Water Program Management

Since 1975, in fact, federal funding for basic state clean water program management grants (under Section 106 of the Clean Water Act) has remained essentially constant while state contributions have increased. The activities funded by these grants form the core of the clean water program, including planning, monitoring, NPDES (National Pollution Drainage Elimination System) permit issuance, compliance and enforcement, and emergency response. Indeed, certain sections of the program having designated funding under the act, such as clean lakes or construction grants management, receive fundamental support from Section 106 grants.

During the past decade, however, while the federal 106 program support level has remained stable, program requirements have grown in scope and complexity. This has been particularly true after passage of the 1977

Clean Water Act amendments. While those amendments—and the initiatives of the senior management (U.S. Environmental Protection Agency)—introduced or emphasized new program requirements (such as toxic pollutant controls, non-point source management, and emergency response), none of the act's previous requirements has been eliminated.

Combined with these increased requirements, inflation has reduced state program grants in real terms. As costs for salaries, fringe benefits, travel and equipment have risen, the federal appropriation for Section 106 has remained essentially unchanged, leading to a serious erosion of resources available to run strong pollution control programs. As noted earlier, while states have partially offset these decreases in recent years by increased support from their legislatures, to some extent, they have also had to streamline their programs. As of this writing many states are discovering that they cannot continue to meet increased requirements, and numerous state legislatures have grown more insistent on limiting the growth of state spending and the size of state agency staffs.

As a result, more and more states are establishing priorities among the act's mandates and attempting to determine which efforts produce the greatest returns in terms of higher water quality. Compliance monitoring, areawide planning implementation, and technical assistance are among the activities most likely to be curtailed in the event of funding shortfalls. An additional result—one that is likely to become more serious in the future when economic activity increases—is a lengthening of the time necessary to process and issue permits under the Clean Water Act (CWA). In states that are subject to mandatory time limits for permit issuance, program managers fear that the quality of the permit reviews will decline.

One system frequently endorsed as a source of additional state revenue for water program administration is the imposition of permit fees. The Office of Management and Budget (OMB) has for many years urged greater reliance on such fees in states with delegated NPDES authority because it envisions the fees supplanting reduced federal program management grants to the states. Unfortunately, of the thirty-four states with NPDES delegation, only a few have fee systems that actually contribute substantially to program support.

Early in 1982, ASIWPCA, in conjunction with the National Governors' Association and the National Association of Air and Solid Waste Officials, took a survey of the states' environmental programs and their funding. The survey specifically sought information on existence of state authority to collect water quality permitting fees, the actual collection of such fees, the recipient of the fee revenues, and the portion of the agency's budget supported by the fees. The findings were quite revealing and they determined that: Some thirty states and territories currently have permit fee authority, but only twenty-two now collect the fees; five others are considering or expecting to impose fees in the near future. Moreover, in only twelve states do the fees collected go to the permitting agency. In the remainder the revenues are deposited in the general fund, and although this avoids the potential conflict-of-interest problem of the levying agency being the beneficiary of the payment and therefore possibly charging unduly high fees, it clearly also precludes or reduces the prospect OMB envisions. That is, state water agencies cannot sustain current federally funded operations by using state fee-generated revenue if they do not receive that revenue. The survey also reveals that, except for New Jersey which obtains 100 percent of the NPDES program budget from its fees, these systems provide from under 1 percent to about 30 percent of state agencies' operating budgets, with the lower level being far more common.

These facts notwithstanding, many states are considering adopting or increasing permit fees. The issue is being examined by water program managers in conjunction with governors, state legislatures, and the affected public. In my opinion, clearly the way in which the water quality program should be funded is a matter of public policy involving economic and political considerations as much as, or even more than, technical ones.

Of course, a number of technical problems can be associated with the development of an equitable fee system. It is possible, for example, that costs of administration can outweigh the income generated. More seriously, the philosophy that "the user pays" can impose a hardship on smaller businesses and municipalities that operate at the margins of the economy. Indeed, it is often the case that smaller businesses have more trouble with their permit applications; if fees are to be proportional to processing costs, it will frequently be the smaller enterprises that will have to pay the higher fees relative to their income.

The political difficulties in obtaining legislative support for the imposition of new costs during a period of economic hardship are self-evident. Even changing fee levels, which commonly necessitates legislative approval, is an arduous task. Despite these problems, more and more states are likely to study permit fees as a source of revenue for their clean water program.

Municipal Wastewater Treatment Facilities

By far the largest financial component of the water quality program for government at all levels is the construction of municipal wastewater treatment facilities. The recognized needs of these facilities have escalated dramatically over the past decade; while $18 billion was committed for the federal share of Publicly Owned Treatment Work (POTW) construction in the 1972 Clean Water Act, EPA officials now estimate that at least $70 billion in needs (in 1980 dollars) will remain after the FY 1985 funds are obligated, at which point about $45 billion in federal funds will have been obligated or expended.

Evaluating Local Financial Capability

Along with the widespread realization that available federal funds cannot approach meeting the needs for municipal facility construction, has come a growing awareness of the responsibilities of state and local governments in contributing to the funding of these plants. Reflecting this new philosophy, in 1981 the EPA's Office of Water Program Operations funded a project designed to assist municipalities in an assessment of their capability to finance the building, operation, and replacement of a POTW. The results of this study were published in the *Financial Capacity Guidebook.*[1]

ASIWPCA welcomed the EPA's increased interest in municipal financial capability, although many of our members are troubled by the regulatory requirement that the financial capability demonstration must accompany the Step 3 (construction) grant application. Specifically, it is ASIWPCA's position that: "Decisions on financial capability and local affordability should be made by U.S. EPA or delegated state agencies during the facility planning phase of the project. . . ." State water program managers simply do not believe that a community can knowledgeably complete the planning and design phases of POTW preparation before analyzing their financial situation. Moreover, they do not believe this determination to be a one-time operation limited to filling in a form; it must be a continuing process throughout the plant's construction and operation.

Despite these reservations, there is much of value in the *Financial Capacity Guidebook*. It is written especially for communities of less than 10,000 people. It therefore speaks directly to those communities where nearly 70 percent of the construction grants are awarded and where historically, public officials have had the least ability to adequately analyze their financial capability. The guidebook gives the officials of such communities a nontechnical "roadmap" to completing a credit analysis that is, in part, comparable to those used by rating services in establishing bond ratings. The analysis measures both the financial condition of the entire community and the financial burden the proposed project will impose on individual households.

Especially useful in light of today's fiscal realities is the guide's emphasis on alternatives that should be considered if the analysis reveals that the proposed conventional treatment system is not financially feasible. These range from redesign of the present system to a reorientation of planning toward smaller facilities or less sophisticated treatment processes to staged project development and to a restructuring of the municipalities' financing.

Several other alternatives focused on what might be called "project avoidance." This includes both innovative and alternative treatment processes (such as overland flow or small-diameter gravity sewers for conveying septic tank liquids to upgrading, rehabilitating, or improving the operation and maintenance of existing systems) aimed at allowing a community to avoid a new project altogether. In much the same way that energy conservation has been touted in recent years as "America's untapped energy source," better operation and utilization of existing wastewater treatment facilities that obviates the need for constructing elaborate new systems may be among the most cost-effective methods for small communities to achieve improved water quality.

It should be noted that the financial advantage now accrued to communities using I&A systems—85 percent federal funding rather than the usual 75 percent—will, under the 1981 amendments, continue beyond 1984. After that date, most POTWs will be eligible for 55 percent federal funds, while I&A projects will still be eligible for 75 percent. This, of course, presupposes that there will be federal funds for POTWs after 1984—a somewhat questionable assumption.

In connection with the 1981 amendments, it should also be pointed out that it is now permitted—and EPA officials are now encouraging—states to provide communities with less than the maximum 75 percent of project costs from federal funds. The premise here is that by reducing the number of federal dollars going to any one project, the number of projects that can receive some partial assistance will be increased. Some half-dozen states have begun using or anticipate adopting such an approach in the near future.

Alternatives to Federal Funding

Frankly, though, these readjustments of the remaining federal dollars for sewage treatment plant construction seem to many state leaders as merely interim coping strategies. A number of states are adopting attitudes similar to that recently expressed by Reginald A. LaRosa of the Vermont Department of Water Resources and Environmental Engineering:

> It appears obvious that the current method of funding water pollution control facilities through federal and state grants is about to terminate or at least be reduced substantially. I fully expect that those officials of state and local government who have relied on us in the past as program managers will again look to us to afford the guidance necessary to re-orient the program's funding.[2]

State water program officials realize that there are likely to be two major outside sources for municipalities to turn to for nonfederal financial assistance: the state and the private sector. Communities can also improve their own financial situation by alterations in their charge systems for facilities or in their general cash management. These latter approaches may in many cases be dependent on state technical assistance to municipal officials.

The State Role

State participation in wastewater facility funding has been an ongoing practice for many years, although it has varied markedly from state to state. In the past, grants have been the most popular form of assistance, with thirty-two states providing direct grant assistance to localities. In most of these cases, the states provide some level of funds to match EPA-provided CWA funds, with levels ranging from under 5 percent to over 30 percent. The higher levels are generally applicable to costs that are ineligible for federal funds.

In twenty states, grants are used to fund projects not receiving any federal money. Often these are projects too low on the priority list to realistically anticipate receiving federal construction grants. The range of state support here is broad, varying from 15 percent to 100 percent of project costs, with the average being 50 percent. In a few states, grant programs are used exclusively for communities not receiving federal funds; in these instances, the maximum level of state assistance is relatively high, ranking from 50 to 80 percent.

The sources of these funds also vary among the states. Most use some type of general obligation bond, and a number supplement general obligation bonds with current revenues. In some states there are tax or royalty revenues specifically designated for wastewater facility grants. Wyoming, for example, funds its grants from mineral impact taxes, while Idaho finances its municipal grants with revenues from inheritance and tobacco taxes.

Thirteen states assist municipal POTWs by providing loans, with four of these also offering local grants. The loan funds are usually available to both federally and non-federally funded projects. Their level in the former case ranges from 10 to 25 percent; in the case of non-federally funded projects, it is commonly 100 percent of project costs.

Again, states obtain the funds for their municipal loans from a variety of sources: general obligation and revenue bonds, current revenues, or set-aside energy tax income. The loans are most frequently secured by local taxes or sewer charges but in some instances are assured against future state aid, such as with revenue sharing funds or by local promissory notes. Under Kentucky law, if a municipality defaults on its loan, the State Pollution Abatement Authority may levy and collect a tax of up to 2 percent on water service purchase. Ohio and California have eminent-domain powers and can seize a facility and impose and collect user charges should a loan not be repaid.

The loan interest rates are generally linked to the states' payments on its bonds, but the methods of calculation vary widely, and in a few states the loans are interest free. In virtually all cases, state loans allow local governments to borrow money at lower rates and with greater repayment flexibility than most of them could obtain if they went directly to the bond market.

A third major form of state financial assistance is provided through bond-financing assistance. In five states—New Hampshire, Vermont, Maine, North Dakota, and Alaska—this takes the form of a bond bank. The banks work by selling their revenue bonds and using the proceeds of that sale to buy the local government bonds. Other states, such as Minnesota, offer bond insurance, which guarantees local debt service payment. The intent of these mechanisms is to use state backing and financial capacity to lower the costs of borrowing for local governments. But because of the caution that must be exercised to avoid local government defaults, these assistance programs often exclude local governments with serious pollution problems if they also have financial problems.

Many states provide assistance to localities above and beyond direct financial aid. In some cases this takes the form of financial technical assistance, including aid in issuing and marketing their local bonds. A number of states offer on-site training on capital and operating financing to local personnel, whereas others have prepared training manuals for local use. A few states subsidize local POTW operating costs to reduce the per-household costs for plant support.

During the past few years, states have explored various mechanisms to enable them to increase their financial aid to municipalities. Loans and loan guarantees appear to many to be the most promising techniques. In some states thought is being given to what amounts to a combination loan-grant program: the provision of low- or no-interest loans from a revolving fund to municipal borrowers, through which the state would in effect be making a grant of interest costs.

In October 1982 New Jersey Governor Thomas Kean announced his plan to establish an "Infrastructure Bank" that would use federal capital grants and state bond proceeds to make low- or no-interest loans to municipalities. The loan approach has substantial advantages, according to the governor, since the bank could provide money for upgrading some two hundred sewerage systems through this means, whereas the upgrading of only eleven systems could be funded by EPA grants.[3]

It should be pointed out, however, that most state efforts to augment available funds require legislative or voter approval. While some states are moving forward with efforts to obtain additional revenue from new bond issuance, others are holding back, believing that the prospects for revenue increases are at present gloomy at best. ASIWPCA members are well aware that the financial problems they now confront are equally serious for other state programs and that the competition for state support will grow increasingly severe.

The Private-Sector Role

In their efforts to grapple with the POTW funding shortfall, state program managers are looking more and more to investment specialists for suggestions. At ASIWPCA's 1982 annual meeting, several speakers making presentations on financial management and new sources of municipal funds elicited great interest from the membership. Pointing out that the traditionally used tax-exempt bond market has suffered under recent economic conditions, and especially under the 1981 Tax Act that indirectly increased the cost of bonds to issuers and reduced their desirability relative to other investments, one Wall Street representative emphasized the need for state officials to tap into the expertise of the investment community for new approaches. Variable-rate bonds, bank letters of credit, and bond insurance were among the variations she suggested to increase the attractiveness of bonds.[4]

In light of the resistance to increased demands on property tax revenues, there has been growing interest recently in enterprise bonds—dedicated to the capital costs of a specific facility—as an alternative to general obligation bonds. The assured revenues for POTWs derived from their user fees can result in very high bond ratings that enhance their investor appeal.

Vendor-assisted financing, under which vendors to the facility provide loans or loan guarantees to the municipality, is another innovative approach that could be applied to POTWs. One of the difficulties here, however, is that this technique is untried in wastewater financing, thus creating some hesitancy among government officials toward its adoption. Similarly, sale- and lease-back transactions, whereby investors invest in the equity rather than the debt of a facility, may be financially attractive but are still untested. In these cases the municipality leases and operates the POTW after its completion.

A representative of a major accounting firm has suggested there is great potential for private-sector ownership and operation of wastewater treatment facilities.[5] Many contend that such ownership would be financially attractive to investors and would result in competitive user fees while meeting environmental requirements. To a great extent, the depreciation provisions of the 1981 Economic Recovery Program Tax Act produce the tax benefits that make such investments particularly attractive at this time.

It has also been suggested that communities using a privatization approach would avoid the large drain on local borrowing power that the existing construction grants program has often caused. In extreme problem-ridden cases, it has been shown that municipalities using such a system could obtain the equivalent of a 50 percent federal grant: in the best cases they might obtain the equivalent of a 70 percent grant.

States are optimistic about the role that privatization can play in POTW construction in the future. Although it would not be usable everywhere—its best chances for success are in communities needing a facility for economic development and willing to lose some control over the project—the approach has elicited great interest in the private sector and certainly aroused marked interest among ASIWPCA members. Many states are also currently considering privatization as an alternative method of funding municipal water clean-up.

The Local Government Role

There are several options local governments can consider to increase their financial capability for wastewater plant construction and operation. One possibility is the imposition or increase of connection fees. These are one-time charges levied on prospective users at the time of connection to the sewerage system. Various questions have arisen regarding such fees, primarily dealing with the equity issues for original versus later users and for costs imposed before and after a plant or sewer system has been expanded.

Another possible funding approach for municipalities is a sinking fund that provides for the gradual accumulation of capital from revenues, which over time yields sufficient funds to replace worn equipment or expand the facility. The particular revenue source for the fund—connection fees, annual user charges, or special surcharges—would have to be determined by the community. Some equity issues also arise in connection with the initiation of such a fund in that some POTW users may question the requirement that they pay for all or a part of the next generation's utilities.

A supportable and sufficient user charge system is another aspect of sound municipal financing. A General Accounting Office report issued late in 1981 discussed the inadequacy of many user charge systems as funding sources for plant operation and maintenance. After studying thirty-six plants in ten states, the GAO found that half of them were not imposing user charges adequate to cover their operation and management costs, and that only three of the plants were setting money aside for plant replacement. Distressingly, many expressed the expectation that they could return "to the Federal construction grants program when replacement or reconstruction becomes necessary."[6] It is essential that communities develop a practical

user charge system as one element of an overall POTW revenue strategy and
that all appropriate costs are equitably distributed among users.

Aside from financial planning directly linked to wastewater facilities,
municipalities can also benefit from general improvements in their financial
management. Many small communities using unsophisticated accounting
systems could substantially defray interest expense and generate interest
income by better cash management. One possibility is to combine the cash
management needs of several adjacent towns in order to gain advantages
both in better paid, more professional personnel and in brighter investment
possibilities.

A key factor here is the necessity to ensure that state laws do not con-
strain the financial options for municipalities. In virtually all states, investment opportunities open to municipalities are limited to high-security
instruments. Here it is appropriate to consider whether any new investment
instruments are available that are not being used or whether state laws
restricting such investments should be altered to allow municipalities higher
yields at risks comparable to those of "old-style" investments.

One example of innovative local financing is the "minibond" introduced in several communities and states over the past four years. This
approach to raising revenues by selling small-denomination, tax-exempt
bonds to local citizens achieved national attention when East Brunswick,
New Jersey, issued $1 million worth of such bonds during a four-month
period in 1978. The idea attracted wide interest, and the following year the
Commonwealth of Massachusetts sold minibonds to finance its water pollution control projects. In 1981 Oregon joined the small group of states that
authorize localities to issue bonds in denominations as low as $100. It is this
type of creative approach that more states will need to consider if their
municipalities are to have the flexibility to attract income from new sources.

The Challenge for the Future

Sewage treatment facilities are only one part of the massive problem of
public works infrastructure deterioration that confronts the nation. Over
the past decade, as escalating inflation diminished the number of capital
projects that could be funded by any community, POTWs enjoyed a relatively favorable status because of the high level and percentage of available
federal funds. Nonetheless, as noted previously, the need for facilities has
far outpaced the available funds. The many reasons for this funding shortfall were thoroughly analyzed by Pat Choate and Susan Walter in *America
in Ruins.*[7] But perhaps Governor Kean summarized them most cogently
when he explained the growing backlog of New Jersey capital needs by say-
ing that "sewers and drainpipes don't make attractive political platforms."[11]

In their book Choate and Walter addressed the necessity for public works investments, pointing out that this "is as essential for national and local economic renewal as investment in our industrial plant itself. Indeed, economic development is dependent upon a sound public infrastructure."[8] This same point was made recently by the director of Tennessee Governor Lamar Alexander's "Safe Growth Team," who stated that in Tennessee they have found the overloaded POTW capacity increasingly limits the state's ability to attract new industry. This fact has contributed to the governor's interest in exploring financing alternatives for plant construction as part of a year-long study on enhancing state growth and development while protecting the quality of life.

The challenges confronting state and local governments in funding clean water over the years ahead will be enormous. Their success in meeting these responsibilities will depend in part on their ability to invent new approaches to problem solving—to anticipate problems and develop management capabilities that emphasize efficiency, productivity, and results. States will have to improve their planning processes so they can better set clear goals, define priorities, and allocate limited resources. New intergovernmental mechanisms are also likely to be needed to achieve coordination, and to resolve conflicts between state governments and their localities, and among adjacent states. Better communication will be essential for intergovernmental coordination and for program justification to the public.

The goal of clean water is too important and too politically popular for the effort not to be made. But success will demand the highest degree of creativity and commitment from all of us.

Notes

1. Government Finance Research Center/Peat, Marwick, Mitchell & Co., *Financial Capability Guidebook,* (Washington, D.C.: GFRC, 1982).

2. Reginald A. LaRosa, Vermont Department of Water Resources and Environmental Engineering, letter to ASIWPCA Executive Director, September 24, 1982.

3. Joseph F. Sullivan, "Kean Seeks Agency to Help Maintain Roads and Sewers," *New York Times,* October 4, 1982, p. A1.

4. Presentation before 1982 ASIWPCA annual meeting, by Karen Gifford, Vice President, Capital Markets Group of Merrill Lynch, New York, N.Y.

5. Presentation before 1982 ASIWPCA annual meeting by Harvey Goldman, Partner, Arthur Young and Company, New York, N.Y.

6. General Accounting Office, "User Charge Revenues for Wastewater

Treatment Plants—Insufficient to Cover Operation and Maintenance,''
CED-82-1, December 2, 1981.

7. Pat Choate and Susan Walter, *America in Ruins: Beyond the Public
Works Pork Barrel* (Washington, D.C.: Council of State Planning Agencies, 1981).

8. Ibid., p. 9.

Funding Clean Water: The Los Angeles County Experience

Charles W. Carry and
Robert P. Miele

Southern California has always placed a high value on water, because it has been an important and integral element in its growth and prosperity. The history of the state is full of important battles for water rights among and within various jurisdictions, both in the courts and at the ballot box. As recently as June 1982, the voters of California turned down an initiative that would have insured construction of the Peripheral Canal, a controversial element in the massive State Water Project. The canal was to be a thirty-year construction program to move large quantities of water from northern to southern California.

On the other hand, the manner in which southern California has dealt with wastewater—the other end of the water supply picture—has generally been rather calm and noncontroversial. However, several events of the last decade have had a dramatic impact on water pollution control in California. The first is the passage by Congress in 1972 of the Clean Water Act, which mandated pollution control efforts unprecedented in the history of this country. National expenditures associated with this program over the past ten years have made it the largest public works project in history. Local agencies responsible for water pollution control have been required to undertake rapid construction programs to upgrade their wastewater treatment facilities while concurrently devising methods to provide local funds, both supplement the money provided by Congress to build these facilities and to operate them once constructed.

The second major event impacting public agencies charged with water pollution control in California was the passage by voters in 1978 of Proposition 13—a property tax reform initiative. Proposition 13 substantially reduced the revenue of sanitation districts. For many agencies, revenue from property tax was reduced by over 65 percent. Coming on the heels of the Clean Water Act, with its requirement for large monetary expenditures, Proposition 13 created an immediate need for sanitation districts to revamp their financial management program in order to continue to meet federal and state requirements and to provide needed services to the public.

This chapter will touch briefly on the institutional relationships between water supply and wastewater management in southern California and also give a brief overview of one agency (the Los Angeles County Sanitation Districts) and its involvement with the Clean Water Act, including "tech-

nical" problems, both current and future. However, the major portion of
the chapter will deal with the financial aspects of providing wastewater
treatment for the residents of the Los Angeles County Sanitation Districts
(some 4 million in number) in response to both the Clean Water Act and
Proposition 13.

Water Supply and Wastewater Treatment in
Los Angeles County

Historically, the task of providing water to southern California has been
dealt with separately from the treatment of sewage in the area. The Metropolitan Water District of southern California (MWD), established in 1928,
has served as the largest wholesaler of water in the state. MWD serves a
population in excess of 12 million people and has a service area of 5,100
square miles. In arid southern California, with an annual average rainfall of
less than 15 inches and no major rivers within a hundred miles, most of the
water must be imported through aqueducts that extend between 200 and 500
miles from the basin. The Colorado River Aqueduct was completed in 1941
to deliver water from the Colorado River. Although the system has historically delivered over 1.2 million acre-feet per year, by 1985 MWD's entitlement will be only 550,000 acre-feet per year. The California Aqueduct, as
part of the State Water Project, began delivering water to southern California in 1975. Approximately 600,000 acre-feet is now being purchased.
Curtailment in Colorado River water deliveries is expected to be offset by
increases in State Water Project deliveries. Local groundwater sources have
been used to supplement imported-water deliveries. Since the early 1950s
these groundwater basins have been overdrafted, leading to adjudication
of the basins. A planned program of groundwater replenishment has been
in operation for a number of years to offset the overdraft conditions.
Replenishment sources include storm water runoff, excess imported water
from both the Colorado River and northern California, and reclaimed
wastewater obtained from treatment of sewage in the southern California
area.
    Reclaimed water used for groundwater replenishment is furnished by
the Los Angeles County Sanitation District. Since 1962 sanitation districts
have planned and implemented a program to reclaim and reuse treated
wastewater. In addition to groundwater replenishment, reclaimed water has
been used for agricultural and urban irrigation, recreational purposes, and
industrial process and cooling water. Reclaimed water has served to supplement the potable water supply of the area. A recently completed study has
estimated that within the next twenty-five to thirty years, reclaimed water
can supply 8–10 percent of the water demands of the area.
    Financing of water supplies in southern California has occurred

through the use of general obligation and revenue bonds, ad valorem taxes, and direct charges for consumer water use. The State Water Project was built by the State Department of Water Resources using statewide general obligation bonds. MWD is responsible for constructing regional distribution systems, while local water agencies provide the facilities to serve their customers. Both MWD and local agencies have financed capital projects through the sale of general obligation and revenue bonds. Further, MWD has used both ad valorem property taxes and direct charges to local suppliers as a means of retiring bonded debt, with operating costs for treatment and distribution facilities being paid entirely from consumer water rates.

Since 1923 wastewater treatment in most of Los Angeles County has been the responsibility of the Los Angeles County Sanitation Districts (LACSD). Established under the Sanitation District Act of the state of California, LACSD is empowered to collect, treat, and dispose of wastewater for the residents and industry of Los Angeles County. The seventy-five cities and unincorporated areas of the county are formed into twenty-four active districts defined by drainage areas. Each district is governed by a board of directors consisting on the mayors on each city within the district. A member of the Los Angeles County Board of Supervisors sits on the board of districts containing unincorporated areas. At present the sanitation districts serve approximately 4 million people, as well as 8,000 industrial and commercial companies in the county. The city of Los Angeles, with a population of 2.5 million, operates a separate system for treatment and disposal of wastewater.

The sanitation districts operate eleven wastewater treatment plants. The largest plant, and the backbone of the system, is the Joint Water Pollution Control Plant, a 365-million-gallon-per-day (mgd) facility that discharges treated wastewater to the Pacific Ocean through pipelines that are approximately 2 miles offshore at a depth of 200 feet. Five additional treatment plants serve the population of the main Los Angeles Basin inland of the ocean. These plants, which range in size from 15 mgd to 65 mgd, provide a high level of wastewater treatment, which allows the effluent from the plant to be reused for irrigation, recreational, and industrial purposes, as well as provide a source of water for groundwater replenishment as mentioned previously. The remaining five treatment plants are smaller, ranging in size from 0.1 to 5.0 mgd and serve outlying portions of the county.

Sanitation District Financing

Historical Perspective

Historically, financing of the sanitation districts was accomplished with general obligation bonds for capital facilities and ad valorem property taxes for operating and maintenance of sewers, pumping plants, and treatment facilities. When a new district was to be formed, a report was prepared detailing the needed facilities and their approximate cost. An election was then held among the residents of the district to form the district. Once the district was formed, a separate election was held to authorize the sale of bonds so that the district could build needed facilities and thus become active. Bonds were paid off by ad valorem property taxes, and a separate tax rate was established each year to provide operating funds for the district. If an area desired to join an existing district, it was allowed to "annex," and an annexation fee was charged. This annexation fee was based on a "buy in" of existing capital facilities, and an additional sum was charged to cover the operating cost until the area could get placed on the sanitation district's tax rolls.

In 1972 the passage of the federal Clean Water Act mandated substantial expenditures of capital by wastewater agencies nationwide to upgrade the quality of wastewater being discharged to our nation's waterways. The timetable established by the act was very severe. To offset this significant and immediate expenditure of local capital funds, the act further provided for federal participation in the financing of local projects to the extent of 75 percent. The state of California, through passage of statewide clean water bonds, provided 12½ percent of capital costs, leaving the local agencies to finance the remaining 12½ percent of the construction costs.

Because of the 87½ percent funding from federal and state sources, the sanitation districts did not have to sell additional bonds to pay their share of the necessary construction required to upgrade treatment facilities. The combination of property taxes and annexation fees was sufficient to pay these costs. However, in 1978 the passage of Proposition 13 mandated that the districts drastically revise its financial program.

Present Financial Program

The sanitation districts' present financial management system combines historical methods of financing with new options that were developed subsequent to the passage of Proposition 13, resulting in a multidimensional program consisting of eight separate components as follows.

Bonded Indebtedness. As described previously, historically, capital facilities were financed by general obligation bonds passed by voters of the district. In those districts that have not yet paid off these bonds, there is still revenue generated for payment of the bonds. However, since passage of Proposition 13, it has become virtually impossible to pass general obligation bonds in the state since it requires approval of two-thirds of all eligible voters. Further, maximum allowable interest rates established by state law make it impractical to sell existing authorized, but unsold bonds.

Ad Valorem Property Taxes. Implementing legislation for Proposition 13 established the rule that all agencies receiving ad valorem taxes prior to Proposition 13 would continue to receive their pro rata share of these taxes. In general, this meant that each agency receiving property tax revenue prior to passage of Proposition 13 had that amount cut by approximately 65 percent. For the sanitation districts, the pre-Proposition 13 income from ad valorem property taxes was approximately $30 million. Subsequent increases in ad valorem taxes resulting from inflation factors allowed by the initiative, as well as sale of property, has increased the districts' revenue from ad valorem taxes to approximately $13 million in the current fiscal year.

Service Charge. Following passage of Proposition 13, the districts participated in a major evaluation of their revenue sources. As a result of changes to the Clean Water Act in 1977, certain new facilities required to upgrade treatment in the system were no longer necessary. Thus the reserves that the districts had been accumulating for the preceding five years to pay for those facilities were freed to be used in the short term to offset the approximate $20 million loss of revenue from ad valorem taxes. Given this reprieve, the districts undertook a six-month study of alternative ways of financing the system. Citizen advisory committees were established to provide input to the process, and public meetings were held in the community to provide further discussion of financing alternatives. At the conclusion of this process, the board of directors adopted the *Revenue Program Report* that established the details of a service charge system. The service charge is based on the contribution of wastewater to the sewage system from various classes of users of the system (residential, commercial, small industrial), and on both the quantity of sewage generated by each user and the quality (strength) of that sewage. A total of thirty-six separate classes were established under the program. A single-family home provides the base unit of sewage contribution to the system, and the contribution from all other user classes is based on that unit (called a *sewage unit*). Thus a high-strength, high-volume user of the system, such as a restaurant, may pay its service charge based on 40 sewage units because it discharges the equivalent wastewater of 40 single-family homes to the sewer. Each year the cost of a sewage unit is established for each district (local sewers and pumping plants peculiar to a given district result in slightly different values of a sewage unit for each district) and is based on the district's capital and operating budget for that year. For fiscal year 1981-1982, the cost of a sewage unit for the large districts ranged from $11 to $24 per year. In the smaller, outlying districts, the cost of a sewage unit ranged from $37 to $96 per year.

The lower sewage unit value of $11 does not represent a typical service charge because there was still some reserve money available in certain districts to help offset the full cost of the service charge in that particular district. Since each district had a different amount of reserves when Proposition 13 passed, not all districts have needed to implement a service charge to date. It is projected that when reserves are depleted from all districts, the full value of the service charge will range from $20 to $25. If the ad valorem tax collected from an average single-family home is converted to a sewage unit cost, it would be approximately $8–$10 per sewage unit. Thus the average single-family home in the district's main system is paying $28–$35 per year for wastewater collection, treatment, and disposal.

The *Revenue Program Report* also evaluated alternative methods of collection of the service charge from districts' customers. The method selected, based on minimal collection cost, was the property tax bill. Since the districts were already collecting ad valorem taxes on the tax bill, it became a relatively simple matter of including an additional line item on all tax bills to collect the service charge.

Industrial Waste Surcharge. The 1972 Clean Water Act mandated that industry had to pay its fair share of both the capital and operating costs of a municipal wastewater system. In response to that mandate, the districts established an industrial waste surcharge program that required each industry on the system to pay for the services provided. Under this program an industry is required to monitor both the quantity and quality of the wastewater it discharges to the district's system. Annually the districts establish factors (costs) for both quantity and quality of wastewater, and the industries use these factors to calculate how much they owe for the sewage services provided by the districts. To simplify the process, the districts have developed several different levels of flow as cutoff points for various types of surcharge programs. All industries that discharge less than 1 million gallons of wastewater per year pay under the service charge program just described. Industries with wastewater flows between 1 and 6 million gallons per year pay a rate based on their flow, which has standard strength factors built into it. These industries can establish their flows based on water usage and losses within the industry. For industries with flows greater than 6 million gallons per year, the districts require that they monitor their flow on a daily basis and that the strength of their discharge also should be monitored periodically. These industries are responsible for calculating their surcharge payment each year based on these measured flows and strength. In 1981 the districts revised the Industrial Waste Surcharge Program to require quarterly, rather than annual, payments of surcharge from large industries.

This program has worked well since it was implemented in 1972. The cost to the districts of implementing the Industrial Waste Surcharge Program is less than 5 percent of the revenue it generates. Periodic audits of industrial companies indicate that the overwhelming majority of them pay their fair share of system cost. The few industries who do not pay their fair share are subject to penalties and interest charges, and on rare occasions the districts have had to resort to the courts to obtain payments from industries.

Annexation/Connection Fees. The annexation fee has been a longstanding revenue source for the districts. However, with the inability to sell bonds as a result of Proposition 13, the districts needed an additional source of revenue to finance future capital facilities. Thus in 1981 a Connection Fee Program was implemented in the districts. The purpose of the connection fee, when coupled with the annexation fee, is to provide for the full replacement cost of sewers, pumping plants, and treatment facilities. The Connection Fee Program requires that each new connection to the districts' sewerage system must pay a connection fee. Further, existing parcels that increase their contribution to the system by more than 25 percent (in terms of flow and strength) must also pay a connection fee. If a new structure replaces a structure that was previously connected to the districts' system, a credit is given. The connection fee rate is developed in a manner similar to the Service Charge Program in that various classes of users are charged for their contribution to the capital cost of the sewerage system. As with the service charge, the single-family home serves as the basic unit of charge (called a *capacity unit*), and other classes pay in proportion to what a single-family home is charged. For the year 1981–1982, connection fee rates ranged from $150 to $500 per capacity unit (that is, per single-family-home equivalent).

Revenue from the connection fee/annexation fee program is put into a special capital improvement fund for use in financing future construction.

Sale of Products. Three by-products of sewage treatment are revenue-producing commodities. As mentioned previously, the final effluent from several of the districts' wastewater treatment facilities is reused for a variety of purposes including replenishment of the groundwater basins. In 1981 the districts sold approximately 10 percent of all their treated effluent. As water supplies in southern California became increasingly scarce, the amount of water reuse will increase. It is projected that, within the next twenty to twenty-five years, approximately 30 percent of the districts' treated wastewater will be reused. The major impediments to additional reuse are the large costs associated with providing distribution systems in a fully developed metropolitan area, along with the cost of removing contaminants to a sufficiently high degree to render the wastewater reusable. (The quality of reclaimed water is very good; however, health concerns preclude its being placed directly into existing potable water distribution systems.) As a result of the previously mentioned study on the long-range role of water reuse in the water supply picture of southern California, MWD is seriously considering funding future reclaimed-water systems.

Throughout the years the districts have negotiated contracts with individual water agencies for the sale of reclaimed water. The price of the water was initially tied to the cost of imported water. However, as imported-water cost escalated rapidly, water agencies were reluctant to pay a comparable price for reclaimed water. Thus the selling price of reclaimed water became a negotiated item and was determined principally on how much a water agency was willing to pay for it. Eventually, the selling price of reclaimed water settled at approximately 20 percent of the operating cost of the wastewater treatment plant, and therefore in recent years the districts have adopted a formula for determining the price of reclaimed water that is equal to 20 percent of the operating cost. For fiscal year 1981–1982, the selling price of reclaimed water using this formula is $35 per million gallons.

The second marketable by-product of wastewater treatment is the gas produced by bacterial degradation of solids contained in the sewage (commonly called *sludge*). This gas, which has approximately 50 percent of the BTU value of natural gas, can either be sold or used within the treatment plant, thus offsetting the cost of more expensive energy. In the past the districts have relied on both of these options. However, with the rising cost of energy, coupled with the increasing energy demands of treatment processes required by the Clean Water Act, the districts are in the process of constructing a power generation facility that will help make the Joint Water Pollution Control Plant (JWPCP) nearly energy self-sufficient.

The third marketable sewage treatment by-product is the processed sludge, which is created as a result of treatment. Once the gas is extracted from the sludge, the sludge is dewatered and then composted to remove additional moisture and also to make it pathogen-free. The composted sludge from the JWPCP is sold as a soil amendment to a local fertilizer manufacturer. This material is sold in both bulk and in bagged products throughout southern California. The districts' contract with the fertilizer company calls for the districts to receive a percentage of the company's gross yearly sales.

Construction Grants. As discussed previously, the Clean Water Act provided for funding of capital projects that it mandated. The districts have made liberal use of this program during the past ten years. Combining past and current projects, it appears that the districts will have received approximately $350 million in federal and state construction grants by the year 1985.

Interest Income. The districts have an active and aggressive program of investing their reserves in a manner to maximize the interest earned and thus reduce the requirement for new revenue from the users of the system. In fiscal year 1981-1982, the districts' investments returned a rate of approximately 15 percent. However, it is apparent that this source of revenue is highly volatile, and therefore, in projecting anticipated income, very conservative estimates of interest income must be used.

Districts' Financial Program

To put the various sources of revenue into perspective, table 6-1 shows a breakdown of the districts' 1981-1982 budget. It can be seen that the major portion of the income is derived from construction grants, ad valorem taxes, service charges, and industrial waste surcharges. The income from reserves is a temporary revenue source and represents money that had been accumulated to build facilities that were subsequently not required. Eventually that source of revenue will disappear. There is no revenue shown from annexation/connection fees in the 1981-1982 budget because the Connection Fee Program was implemented in 1981, and therefore, no money was in that fund when the budget was prepared. Prior to 1981, annexation fee income was not accounted for separately, and thus it appears as part of the reserves. Subsequent districts' budgets will contain a line item for annexation/connection fee income.

Table 6-1
Los Angeles County, Sanitation Districts, 1981-1982 Budget

	Millions of Dollars	*Percent of Total*
Income		
Construction grants	48,300,000	45
Property tax	12,800,000	12
Service charge	15,400,000	14
Industrial waste surcharge	16,400,000	16
Interest	5,700,000	5
Sale of products	2,300,000	2
Reserves	6,300,000	6
Total	107,200,000	100%
Expenditures		
Capital facilities	62,800,000	59
Administration	4,500,000	4
Technical support	7,600,000	7
Treatment plants O&M	27,800,000	26
Sewers and pumping plants O&M	4,500,000	4
Total	107,200,000	100%

Potential Future Problems

The 1972 Clean Water Act represented such a major change in the direction of environmental law in this country that it naturally created a multitude of problems at both the national and local levels. It soon became apparent that the timetables established by the act were totally unrealistic. Neither federal nor local agencies were adequately staffed or financed to meet deadlines for construction of wastewater treatment facilities. To further complicate the issue, there was a wide divergence of opinion among the professionals in the wastewater profession about the need for the programs mandated by the Clean Water Act. While everyone concurred on the need to improve our nation's waterways, there was vast disagreement with the act's philosophy, which required uniform treatment nationwide without regard to local water conditions.

Subsequent changes to the act have mitigated some of its earlier controversial postures, and, in general, there is much greater acceptance of the revised Clean Water Act in the profession. However, there still appears to be several areas of concern with the act that link the technical and financial aspects of clean water. In the technical area, the Clean Water Act still requires that all industries provide the same level of pollution control (pretreatment) before discharging to a municipal system, regardless of the geographical location of that industry and the local receiving-water conditions. While this requirement has only a minimal direct impact on local wastewater agencies, it could have a substantial impact on consumers in general if industry has to install costly pollution control equipment. A second technical issue that appears to remain unsolved is the disposal of the residues (sludge) of sewage treatment. The Clean Water Act's mandate for improved sewage treatment has resulted in greater quantities of these residues that require disposal. In coastal communities the deep ocean heretofore provided the disposal mechanism for this sludge, but this method is currently outlawed by the Clean Water Act. Many communities have turned to incineration as a disposal mechanism, but increasingly stringent air quality standards are either prohibiting this option or rendering it financially infeasible. The third option—land disposal—is also threatened by the increasing unavailability of landfill space nationwide, along with concerns regarding the impact on groundwater and crops from contaminants contained in sludge used as a soil amendment.

From a financial perspective, the cessation of federal funding under the Clean Water Act is seen as occurring as early as 1985. Clearly, not all the facilities mandated by the act will be constructed by that date, and thus many municipalities will be left in a financial dilemma. The existence of federal and state funds for construction has often lulled local agencies into a false sense of financial security and dissuaded them from developing long-term, local revenue sources. Thus an immediate cessation of external funds will force these local agencies to further delay the needed capital facilities until adequate local sources of revenue are developed. This in turn will result in an even higher price tag for the nation's clean water program as inflation drives the cost up.

Summary

Historically, water supply and wastewater disposal have remained separate entities in southern California. Both have been handled by large regional agencies. Water supply has been and remains a matter of great controversy in the area. Financing of water supply has occurred at both the state and local level. Wastewater treatment has, until the recent past, been a low-profile industry in southern California. It has had relatively little public exposure and has been financed locally. Starting with the Clean Water Act of 1972, wastewater treatment has been elevated in local exposure and has been financed at both the federal and state level, as well as with local funds. The 1978 passage of Proposition 13 in California required the Los Angeles County Sanitation Districts to diversify its financial program to cope with the mandates of the proposition, as well as to continue to finance facilities required by the Clean Water Act. This new program should also provide a sufficiently strong financial base to withstand the rapidly appearing sunset of federal funds.

7

Funding Clean Water in Illinois

Richard J. Carlson and
Roger A. Kanerva

Water Supply and Quality

Historically, Illinois has been considered a water-rich state—with good reason. It is, in a sense, surrounded by fresh water, with the Mississippi River serving as its western border, the Ohio and Wabash Rivers to the south and east, and Lake Michigan to the northeast. Within its interior the large tributaries to these major river systems are supplemented by nearly 83,000 inland lakes and ponds, most of which are manmade. Lake Michigan is the single most important body of water available to the state, supplying water to Chicago and numerous suburbs, as well as providing navigation and recreation.

Average annual precipitation in Illinois ranges from 32 to 46 inches per year. Heavy rainstorms that produce floods may occur at any time, but their frequency is greatest from late spring to early fall. Droughts have been observed in all sections of Illinois and are most frequent in the south and southwest.

Groundwater is used in every county and is withdrawn from fourteen major aquifers. The potential for groundwater development in Illinois is substantial, on the order of 7 billion gallons per day. For the most part, present withdrawals are far less, averaging about 1 billion gallons per day. However, in the Chicago metropolitan area, pumpage from deep sandstone wells is three times the rate of recharge, and water levels are declining rapidly. There are also a few small public water supplies dependent on local aquifers with limited recharge capability.

Water Uses

According to the Illinois State Water Survey,[1] water withdrawals from all sources in the state exceeded 42 billion gallons per day in 1980. The electric power generation industry accounts for 94 percent of all water use, but more than 99 percent is returned to its source with only a slight increase in temperature. If the electric power industry is excluded, the remaining water use is accounted for by public water supplies (67 percent); manufacturing (18 percent); rural uses (11 percent); and mineral extraction (4 percent).

87

Public water supplies furnish nearly 95 percent of the state's population with potable water. Surface water supplies about 64 percent of this total, and groundwater supplies 46 percent. This leaves 575,000 people who furnish their own water. The largest system in terms of people served is the Chicago Department of Water, serving more than 4,500,000. The largest geographical area served by a public water system is the Rend Lake Conservancy District, which serves parts of nine counties stretching over more than 1,800 square miles in southern Illinois.

Three major industries account for nearly two-thirds of the total used for manufacturing primary metals, chemicals, and food products. Rural water withdrawals currently constitute one of the fastest growing uses of water in Illinois.

Wastewater Treatment Facilities

Illinois currently has 772 publicly owned wastewater treatment works (POTWs) and an estimated 35,000 miles of sewers. The IEPA (Illinois Environmental Protection Agency) estimates the value of these facilities to be about $11.4 billion. Nearly 97 percent of these POTWs provide the equivalent of secondary or better treatment, meaning that over 70 percent of the state's population is currently served by treatment facilities that meet basic pollution control standards. Of the remaining population, about 26 percent are served by septic systems, about 2 percent by primary treatment facilities, and about 2 percent by miscellaneous private systems.

Adequacy

Two elements are relevant in assessing the adequacy of current wastewater facilities and the need for additional investment: (1) compliance with state and federal water pollution control requirements and (2) the capacity available to accommodate growth.

Federal and state water quality goals, reflecting the public's desire for environmental quality, are the driving forces behind the nation's investment in wastewater treatment facilities. The goals set out in the federal Clean Water Act—fishable, swimmable waters by 1983 and zero discharge of pollutants by 1985—are proving to be unrealistic, but they clearly represent the kind of environment in which most Americans would like to live. Any assessment of the adequacy of current facilities and the need for additional investment must take this political and social dimension into account. A significant shift in popular attitudes could mean a significant shift in regulatory programs, especially as the cost of zero discharge goals becomes evident and as local communities must increasingly bear the cost of improvements.

The common yardstick for measuring the costs of compliance with state and federal regulatory requirements is the biennial survey published by the U.S. EPA entitled "Cost Estimates for Construction of Publicly Owned Wastewater Treatment Facilities." First conducted in 1973, the needs survey is an attempt by the EPA to develop state-by-state and nationwide estimates of the costs of complying with the Clean Water Act.

Under the act all municipal sewage treatment plants are to achieve secondary treatment by 1988. [Secondary treatment is a technology-based performance standard that is defined by the EPA. In general, it provides for 85 percent removal of biochemical oxygen demand (BOD) and suspended solids.] This requirement represents a national treatment goal for municipal facilities. However, where secondary treatment will not maintain the water quality standards of the receiving waters, a higher degree of treatment must be provided. Water quality standards are set by each state to reflect the natural characteristics and human uses of that state's rivers and lakes and must be approved by the EPA and be consistent with federal guidelines.

The federal construction grants program is designed to assist municipalities in meeting the treatment goals of the Clean Water Act, which provides grants for five categories of treatment. The 1982 needs survey estimated the national cost of constructing wastewater treatment facilities to serve year 2000 populations to be $118 billion for all categories. The cost estimates for facilities in Illinois are listed as follows in millions of 1982 dollars.

	Needs Category	Estimated Costs	Percent of Total State Needs
I.	Secondary treatment	$598	13.1
II.	Advanced treatment	965	21.1
III.	Major sewer rehabilitation and correction of infiltration inflow	233	5.1
IV.	New collector and interceptor sewers	677	14.8
V.	Correction of combined sewer overflows	2,096	45.9
	Total	$4,569	100.0

The IEPA feels the needs survey overestimates the true treatment needs of the state in light of regulatory changes now pending at the state and national levels and the ultimate affordability of meeting all applicable standards.

The Illinois situation varies in significant respects from the national pattern. Advanced treatment accounts for over 21 percent of the state's needs but only around 5 percent of national needs. This stems from the state's strong commitment to comprehensive water quality standards and control of toxic pollutants—a commitment that often translates directly into advanced treatment requirements. These include numerical water quality standards for a wide variety of pollutants, including ammonia, and effluent limits for BOD and suspended solids that are more stringent than those included in the current EPA definition of secondary treatment.

Illinois, like many other industrialized states in the Midwest and Northeast, has many communities with combined sewers—those that carry both sewage and stormwater to the treatment plant. During storms, these systems may carry flows that exceed the capacity of the treatment plant. The excess flow, including raw sewage, is typically discharged directly into a stream without treatment. Combined systems not only contribute to pollution problems but also cause flooding and basement backups during wet weather. Over 50 percent of the people in Illinois are served by combined sewers, many of them in the Chicago metropolitan area. Combined sewer needs amount to almost half of the total state's needs in the 1982 survey but only about 30 percent of the national needs. While the EPA estimate is probably much too high, the discrepancy highlights how needs will vary from state to state.

There are still 357 wastewater facilities in Illinois not meeting final effluent limitations in their permits. Most communities not in compliance are pursuing projects under either the state or federal construction grants programs. While the rate of progress in upgrading facilities has slowed due to reductions in federal and state funding, the IEPA is optimistic about the amount of facility improvement expected in the next four years.

Population growth and shifts in the distribution of population have a direct bearing on the adequacy of the state's wastewater management system. Statewide, the average reserve capacity for additional wastewater flows (excluding the Metropolitan Sanitary District of Greater Chicago) varies from nine years in suburban Chicago to twelve years in the rest of the state. The availability of this much reserve capacity is a result of the continuing investment made over the past twenty years. This achievement is even more significant when one considers that Illinois' population increased 16 percent over that time period and that the state experienced extensive industrial growth as well.

Statewide averages, however, can be misleading. Growth has not taken place uniformly across the state, and there have been significant shifts in the distribution of existing population (that is, from major urban centers to suburban communities). As a result, the IEPA has had to place fifteen communities on restricted status (zero remaining capacity) and sixteen communities on critical review (less than 20 percent remaining capacity). Many of these communities are currently pursuing grant projects to remedy capacity problems. In the meantime there remain pressing capacity problems in a dozen or so communities. In several instances, however, the slowdown in the economy has allowed some of these communities to catch up with anticipated growth. At the other extreme there are a few communities with reserve capacity for the next two decades due to extensive industrial shutdowns and modest population growth.

An assessment of capacity must also take into account the time required to complete new wastewater projects. In effect, project implementation time serves to discount the available reserve capacity. An average project takes 5.5 years to complete under the federal and state grant programs. However, local governments will be funding most of the next round of improvements on their own. Since such projects could probably be completed in a more timely fashion without the constraints of the grant programs, it may be that three years represents a reasonable project implementation time. When discounted by three years, capacity estimates reveal only six to nine years of investment-free use of facilities. In other words, some communities in Illinois will soon be facing another round of public works investments.

Assessing the adequacy of facilities remains somewhat elusive due to changes occurring in the federal clean water program. For example, the EPA is currently considering a new definition of secondary treatment that could ease requirements for some facilities. The EPA is also proposing major changes in the procedure for setting water quality standards that are expected to provide relief for some communities. As a result, it will be several years before the adequacy of the state's facilities can be fully assessed.

Maintenance

The operation and maintenance of wastewater facilities have a direct effect on their ability to meet standards and accommodate growth. Poor maintenance can lead to rapid deterioration and poor performance. Proper operation and maintenance require technically competent management, adequate staffing levels, and a commitment by the local government to generate sufficient revenues to replace worn-out equipment. The IEPA has developed an active operator certification program that is enforced through its National Pollutant Discharge Elimination System (NPDES) permits for POTWs, since properly trained operators enhance the prospects for sound maintenance and operation. Facility maintenance is a local responsibility, and until very recently, the IEPA did not attempt to track local funding for maintenance. However, starting in 1981, NPDES permittees are being required to submit annual reports regarding user charges, operations and maintenance funding, and facility staffing. Once all the permits have been revised to require these reports, reliable information will be available to accurately assess the adequacy of local management.

In 1981 the IEPA did a special study of user charges in three hundred POTWs that had been expanded or upgraded since 1970. The average monthly user charge was $8.08 per household. Wastewater collection and treatment was costing the users of these facilities about 10¢ per person per day. While this analysis did not include charges relating to ad valorem taxes, the adjusted figure should not be significantly greater than the $8.08 level.

Of course, the charges for individual communities varied relative to economies of scale, location, and other factors. Actual user charges ranged from $5.00 to $12.00 per month. During the review of facilities plans, the IEPA normally reviews the extent to which a community can afford a project, particularly if the proposed monthly charge exceeds $20 per household. In a number of instances, grantees have been advised to reconsider their projects to prevent a financial hardship on the community.

Keeping costs at a reasonable level must, however, be balanced against the need for adequate facility maintenance. The majority of POTWs in Illinois are probably providing an adequate level of operation and maintenance funding, particularly where communities have adopted user-charge ordinances consistent with EPA regulations. Indeed, in several communities user-charge revenues are running well ahead of expenses. Others, however, have failed to update their user charges and have allowed their facilities to deteriorate beyond acceptable performance levels. In many of these latter instances, operation and maintenance costs were subsidized with tap-on fees from new development. When growth declined, these communities had large operation and maintenance deficits and little time to correct the situation. Eventually, the IEPA was forced to pursue enforcement actions to bring these communities back to a sound operation and maintenance program.

Results

The results of investing in wastewater treatment facilities must be measured in terms of compliance with regulatory requirements and measureable improvements in water quality. There are 753 treatment works with NPDES permits in Illinois. Of these, 399 have been expanded or upgraded since 1970 with federal or state grant funds. These facilities range from simple lagoon systems to highly complex activated sludge systems with nitrification and filtration. One way to assess compliance is to analyze the extent to which these facilities are meeting the effluent limitations set forth in their permits for conventional pollutants such as BOD and suspended solids. [The EPA definition of secondary treatment calls for no more than 30 milligrams per liter (mg/l) of either BOD or suspended solids in the discharge from a treatment plant (30/30).] However, more stringent limits are often needed to meet the state's water quality standards.

Determining compliance with these standards is complicated by the application of interim and final effluent limits. When the NPDES permit program was first implemented in the mid-1970s, many POTWs were unable to meet the effluent limits needed to satisfy either the EPA secondary treatment definition or state water quality standards. Initial permits were written that acknowledged these limitations by setting interim effluent limits that these plants could meet, which in some cases were substantially below EPA standards for BOD and suspended solids. The permits also contained compliance schedules for ultimately meeting the final effluent limits necessary to meet federal and state requirements. As a practical matter, adherence to these schedules depended on the availability of grant funds. As a result, however, compliance with interim limits is substantially different than compliance with final limits in terms of water quality impacts. Based on a plant's ability to meet applicable effluent limits for each month of the year, the 1982 compliance rate for facilities with final effluent limits in effect was 94 percent. For those subject to interim limits, the compliance rate was 84 percent. Over 67 percent of all facilities had no violations for the entire year.

Even more revealing is the overall performance of plants built after the advent of the major federal grants program as compared to those built earlier. The average quality of the effluent discharged by all plants built before 1970 is 35/43, exceeding even the secondary treatment standards. For those built after 1970, the average effluent quality is 17/28. Plants built in conformity with the federal and state grant programs clearly perform much better than the state's older POTWs.

All of the state's 753 POTWs with NPDES permits should eventually be subject to final effluent limits that will enable them to meet the state's water quality standards. Meeting this goal is contingent on the availability of federal, state, and local funds for upgrading facilities. The following table summarizes the effluent limits now required in Illionis, the number of POTWs currently required to meet them, and the number that must ultimately comply.

Final Effluent Limits	POTWs Currently Subject to Final Effluent Limits	POTWs that Must Ultimately Comply	Total
10/12	162	238	400
20/25	45	19	64
30/30	175	114	289
Total	382	371	753

Compliance with regulatory requirements has contributed to a growing improvement in water quality. Since 1958 the state has conducted an ambient stream-monitoring program to characterize the physical, chemical, and biological condition of the state's surface waters. The network now consists of 204 stream stations, and deviations are measured against the water quality standards adopted by the Illinois Pollution Control Board.

Historical comparisons are complicated somewhat by changes in the nature and location of sampling instrumentation; yet where such comparisons can be made, there have been some significant improvements. Comparisons of 1980–1981 data with 1978–1979 data show the average concentration of total phosphorus was reduced by 24 percent, with 57 percent of the monitoring stations showing a decrease. Ammonia was reduced by 8 percent, with 53 percent of the stations showing a decrease in concentrations. For the 1972–1973 through 1980–1981 period, the average ammonia concentration decreased by 33 percent.

While significant gains have been made, much remains to be done. The Illinois Water Quality Inventory Report for 1982 shows little overall improvement in several water quality parameters over the past three years. The reason for this apparent lack of overall improvement is twofold. First, past efforts to control conventional pollutants from point sources have reached a plateau in terms of water quality. We have seen the gains attributable to the first round of major investment in upgrading facilities. However, several hundred publicly owned treatment works are still in the process of completing design or construction of new facilities. (Many of these second-generation POTWs will utilize advanced treatment technologies that will further reduce discharges of pollutants such as ammonia.) In addition, the first phase of the major combined sewer overflow program in the Chicago area is not scheduled for completion until 1985. Thus in the next few years there should be new improvements in water quality resulting from the most recent efforts to control point sources of pollution. Second, the state's remaining water quality problems are largely attributable to nonpoint sources—those that do not discharge from a pipe or outfall. Examples of nonpoint pollution include runoff from agricultural and mining activities and urban stormwater drainage. The problem of nonpoint source pollution is just beginning to be addressed in Illinois. The water quality management planning process, mandated under Section 208 of the Clean Water Act and completed in 1979, identified the major sources of nonpoint pollution in the state and the management strategies to address them. A program to encourage management practices to control agricultural soil erosion is being implemented by the Illinois Department of Agriculture, and resulting gains in water quality improvement should become evident in the next few years.

Public Water Supply Facilities

Most people in Illinois obtain their domestic water from "community" public water supply systems, those which serve at least twenty-five residents. The remaining systems, including households and businesses, obtain water from private wells or, in the case of some larger industries, from their own surface impoundments. There are 1,992 public water systems in the state with roughly 60,000 miles of water mains. Over 90 percent of the supplies (1,792) serve communities with populations under 10,000. Most rely on groundwater and are publicly owned and operated.

Unlike wastewater treatment facilities, the investment in public water supplies has been driven primarily by service needs rather than compliance with state and federal environmental regulations. It has been the amount of water necessary to meet community and industrial needs, rather than its quality, that has governed the development of new water supplies or the expansion of existing ones. Recently, however, there has been a growing emphasis on the quality of drinking water. Full implementation of the federal Safe Drinking Water Act will mean increased expenditures to test for and remove nonconventional contaminants such as trihalomethanes and organics. The public is growing more aware of and concerned about the presence of toxic substances in ground and surface water and public water supply officials will be under increasing pressure to treat for these substances. This means, of course, that considerably more money must be spent on treatment systems.

Public water supplies are different from wastewater treatment facilities in another important respect: There has never been a major federal or state grant program to assist local communities with the expansion or development of drinking water facilities. Aside from modest federal programs to provide grants or loans to small rural communities, local governments and private water utilities have had to finance construction and operation from local resources. The absence of a financial assistance program means that the state has not had the same opportunity to develop accurate information on the need for future investments as it has had for wastewater treatment systems.

Public water supplies do share one important, and relevant, characteristic with sewage treatment systems: a tendency for local officials to underprice their service, often charging rates too low to cover routine maintenance, much less major repairs and rehabilitation. The real cost of drinking water is often disguised by deferring needed capital investments.

Adequacy

Public water supply systems in Illinois are generally meeting the needs of the domestic and industrial users that they serve. However, there are a few isolated communities in the state where the quantity or quality of ground and surface water is or soon will be a problem. Sediment buildup in surface water supplies is a major headache for certain supplies in central and southern Illinois. The Illinois 208 Water Quality Management Plan has identified soil erosion from all sources at more than 181 million tons annually. Erosion from cropland contributes more than 86 percent of the total. Reservoirs within the state lose some 8,000 acre feet of storage capacity each year due to sedimentation (which would cost an estimated $26 million to remove by dredging). Sedimentation also causes damage to fish and aquatic organisms and increases the cost of treatment for public water supply systems. For a few supplies in southern Illinois, low rainfall can mean the rapid depletion of existing reservoirs.

Recent studies of public and industrial water supplies in the Chicago metropolitan area have shown that groundwater withdrawals from the deep sandstone aquifer underlying the region exceed by three times the natural recharge rate. The major alternative to groundwater withdrawals is the use of Lake Michigan water; however, withdrawals from the lake are limited by a Supreme Court consent decree to 3,200 cubic feet per second (cfs). Currently the city of Chicago and 111 suburban communities rely on Lake Michigan as their source of raw water. The consent decree was amended in 1981 to change the method of accounting for withdrawals and to place restrictions on the use of the water for diluting wastewater. The change freed up approximately 150 cfs for domestic water supply uses to be allocated by the Illinois Department of Transportation (IDOT). Through a public hearing process, eighty-five new users received allocations—but to gain access to the water, they have to invest substantial amounts of money in hookups to existing treatment and distribution systems. For example, four northwest suburban communities plan on spending $90 million to tap into an existing north suburban system that draws on Lake Michigan water. This project cost is typical of the level of public investment the new users will face in making use of their allocations. It should be added that these projects should substantially reduce current groundwater withdrawal rates but will not eliminate the mining of the aquifer.[2]

One of the conditions that new Lake Michigan water users were required to meet was the imposition of a local water conservation program. This requirement not only reflected the practical necessity of maximizing the allocation of water but also the state's growing involvement in water conservation measures. For the last several years the Illinois Department of Commerce and Community Affairs has had an active technical assistance program to help localities make the best use of their existing water supplies through sound management practices and conservation techniques. Assistance is also provided to individuals and public-service groups through field contacts, workshops, and publications. The department also assists communities in obtaining loans and grants needed to improve water supply and treatment systems.

Recently the IDOT made a rough assessment of the adequacy of major surface and groundwater systems for year 2000 populations, focusing on the raw water source as well as treatment and storage capacity. In its report IDOT stated that out of 691 public groundwater systems: 5 percent (38) had marginal or deficient supplies; 6 percent (41) had marginal or deficient treatment works; and 58 percent (401) had marginal or deficient storage capacity. These figures suggest that sufficient groundwater should be readily available to most, if not all, existing supplies for the next two decades, with relatively modest investments in the expansion of existing treatment and storage facilities.

Currently the IEPA has placed 205 public water supplies on restricted status, which means that the IEPA will not issue permits for water main extensions except under certain very limited conditions. Restricted status results from persistent violations of drinking water standards or deficiencies in equipment that could result in violations. Correction of these problems often requires some expenditure of funds by the local government.

Maintenance

Adequate maintenance of public water supply facilities is critical to their performance and ability to meet demands for growth. The age of facilities in Illinois vary greatly. In areas experiencing recent growth, like the highly urbanized northeastern portion of the state, facilities tend to be new or recently expanded. The older facilities are found primarily in small towns and rural areas. Age is a factor, but the condition of a plant often depends more on maintenance practices. There is an unfortunate reluctance on the part of many communities to set aside the revenues needed to properly maintain and upgrade their facilities. The users of public water supplies, like the users of wastewater treatment plants, tend to take these public services for granted and resist paying user charges at levels needed to maintain the equipment.

Results

Drinking water in Illinois is generally safe, clean, and adequate in quantity. More than forty years have passed since any communities in the state experienced significant epidemics relating to waterborne disease. Episodes still occur on occasion, the latest being in 1981, but these usually only result in gastric enteritis, an uncomfortable but self-limiting group of diseases with usually no long-term effects.

Funding: History, Problems, and Alternatives

The state program to protect public water supplies began in 1915 with the establishment of the Division of Sanitary Engineering in the State Board of Health. This was followed in the early 1920s with the establishment of mandatory standards to protect drinking water. Statewide efforts to control water pollution began in 1929 with the creation of the Sanitary Water Board, which was initially responsible for assisting municipalities and industries in complying with water quality standards and regulations. In 1970, passage of the Illinois Environmental Protection Act transferred permitting and enforcement responsibilities for water pollution and public water supplies to the newly created Illinois Environmental Protection Agency (IEPA) and rule-making authority to the Illinois Pollution Control Board.

Grants for the construction of sewage treatment facilities in Illinois first became available in 1956 when Congress passed the federal Water Pollution Control Act. This modest program was expanded in 1966 when amendments to the act raised the federal share for a project to 55 percent and eliminated the ceiling on project costs, allowing cities of all sizes to participate in the program.

In November of 1970 the voters of Illinois approved a $750 million general obligation bond issue to finance the "planning, financing and construction of municipal sewage treatment works." Funds from the sale of these bonds were initially used to match federal grants. This generally provided project cost sharing of 55 percent federal, 25 percent state, and 20 percent local for grant-eligible costs. This pattern continued until passage of the federal Water Pollution Control Act Amendment of 1972 (Clean Water Act), which increased the federal share ratio to 75 percent. In 1973 the state grant program was changed in response to the new federal amendments. Since it was federal policy not to award grants at percentages lower than 75 percent, the state decided to use its resources to establish a separate but complementary program of 75 percent state grants that would also require a 25 percent local share. The two programs were coordinated through the use of the same federally approved priority system and grant requirements to determine project eligibility. Both federal and state grant programs consist of a three-stage process. Eligible applicants (that is, local governments including sanitary districts) first undertake a facilities planning process to identify specific water pollution problems and alternatives to solving them (Step 1). Once a cost-effective solution is chosen and approved by the Agency, the applicant moves into design (Step 2). When the design is completed it must be approved by the Agency through the issuance of a construction permit. Then the applicant can begin construction (Step 3). This procedure served to prevent grant shopping by applicants. The scope of the state program was expanded in 1975 to allow the funding of projects outside the normal priority range in order to encourage the regionalization of collection and treatment systems and to address health hazards.

The state and federal construction grants programs have been the foundation of the municipal facility compliance strategy in Illinois, and they continue to occupy a very high priority in the state's water pollution control program. The availability of financial assistance has been critical in upgrading a wide range of municipal treatment systems and in maintaining the state's progress in improving water quality. The key role played by the grant programs has led the IEPA to develop strong working relationships with local government officials, the consulting engineering community, and the EPA. These relationships continue to improve even in the face of ever-changing requirements, whether they are imposed administratively by the EPA or legislatively by Congress.

In Illinois there are 908 geographic areas designated by the IEPA to perform facilities planning. These areas vary from complex urban locations to individual rural villages with populations ranging from 150 to over 5 million people. The vast majority of these areas are currently served to some extent by sewers and treatment facilities. As of December 1982, 774 applicants had received facilities planning (Step 1) grants, leaving a balance of 134 applicants yet to start the planning process. Most of the areas that have not begun facilities planning are small, unsewered rural towns that may never need central collection and treatment facilities or whose facilities are unlikely to need expansion or upgrading.

To date, 449 applicants have completed Step 1 requirements and their facility plans have been approved by both the IEPA and the EPA. Based on the IEPA's experience, nearly 250 of the 908 facility planning areas will conclude that no water pollution abatement needs exist in their service area upon completion of planning. There have been 514 design (Step 2) grants awarded from state or federal funds, and 465 applicants have completed Step 2 requirements. Of those applicants now in Step 2, all will be allowed to begin construction as soon after approval of Step 2 work as the availability of funds allow. Finally, 514 construction (Step 3) grants have been

Table 7-1

Federal and State Grants: Appropriations and Obligations, 1973–1983
(thousands of dollars)

Fiscal Year	Appropriations			Obligations		
	Federal	*State*	*Total*	*Federal*	*State*	*Total*
1973	124.9	33.1	158.0	80.4	33.1	113.5
1974	187.5	0	187.5	53.0	0	53.0
1975	252.3	170.5	422.8	362.4	170.5	532.9
1976	571.7	130.4	702.1	337.4	130.4	467.8
1977	52.2	73.9	126.1	325.5	73.9	399.4
1978	237.7	46.1	283.8	264.3	46.1	310.4
1979	215.1	41.4	256.5	220.4	41.4	261.8
1980	150.5	35.3	185.8	150.5	35.3	185.8
1981	130.5	35.0	165.5	98.1	35.0	133.1
1982	122.7	28.4	151.1	18.2	17.9	36.1
1983	110.7	0	110.7	67.9	1.3	69.2
Total	2,155.8	594.1	2,749.9	1,978.1	584.9	2,563.0

awarded from state or federal funds since the 1972 amendments were passed, and 282 of these projects have been completed and are in operation today. The rest are under construction.

Table 7-1 indicates the total amount of federal and state funds appropriated and obligated for fiscal years 1973 through 1983. Table 7-2 includes the number of grants issued by step and by fiscal year.

Future Needs

The state must face numerous programmatic issues if it wishes to continue making progress in improving water quality and ensuring the adequacy of drinking water. A few of the most significant issues are discussed in the following paragraphs.

Federal Grants

A principal concern of the water pollution control program is the continuation of federal funding for municipal sewage treatment construction grants. If federal funds for construction grants are phased out in the near future or if the level of appropriation is substantially reduced, the water pollution

Table 7-2
Number of Federal and State Grants Awarded, 1973–1983

Fiscal Year	Step 1		Step 2		Step 3		Combined[a]		Total
	Federal	State	Federal	State	Federal	State	Federal	State	
1973	0	0	0	0	25	0	14	0	39
1974	3	0	0	0	16	0	1	0	20
1975	84	56	8	8	21	22	0	70	269
1976	168	116	23	1	27	0	0	86	421
1977	99	50	37	7	30	1	0	33	257
1978	50	8	30	6	30	0	0	30	154
1979	62	2	14	0	14	3	0	31	126
1980	37	2	22	3	13	0	11	16	104
1981	32	5	17	0	12	1	2	15	84
1982	16	3	20	0	8	3	6	10	66
1983	5	0	9	0	18	0	14	0	46
Total	556	242	180	25	224	30	48	291	1,586

[a]This category refers to grants that combine design, planning, and construction into a single grant.

program will be faced with a difficult situation. How will sewage treatment construction needs be financed? Will the state or individual localities provide the financing for sewage treatment facilities necessary to comply with existing water pollution regulations? How will the cost of constructing additional sewage treatment facilities be factored into a review of design standards and regulations?

Some reduction in the level of federal support is inevitable. The Clean Water Act was amended in 1981 to reduce the federal share ratio from 75 to 55 percent and to eliminate certain categories of funding (that is, major sewer rehabilitation and new collector sewers). Congress passed a four-year authorization and has appropriated $2.4 billion in FY 1982 and FY 1983. The amendments do not take effect until October 1, 1984 to allow states a period of transition to the new system. Funding beyond federal fiscal year 1984 is uncertain. A significant level of federal funding will continue to be needed to meet federal and state water quality goals, however, even if state and local governments substantially increase their level of investment.

As indicated earlier, the EPA 1982 needs survey estimates the total remaining needs for wastewater facilities in Illinois to be $4.569 billion. This estimate is too high, particularly with respect to treatment needs for combined sewer overflows. While possible changes in both the federal and state programs complicate the picture, a more accurate estimate of total need is $1.75 billion. These wastewater facility needs will be funded from federal construction grants appropriations, state bond funds, and local financing. Available and anticipated federal funds, along with the associated local-share requirements, are illustrated in the following tabulation (in millions of dollars):

Fiscal Year	Federal	Local	Total
1982	115	38.3	153.3
1983	110	36.7	146.7
1984	110	36.7	146.7
1985	110	90.0	200.0
Total	445	201.7	646.7

Based on these projections, there will probably be slightly over $1 billion in treatment needs to be funded in 1986 and subsequent years.

Only about $75 million in state bond funds remain for obligation in 1984 and beyond. However, new funds may be made available sometime in the future by the general assembly. If federal grant support continues at current levels and state and local governments are able to market sufficient bonds, it is conceivable that almost all needs could be satisfied with another ten years of federal, state, and local funding at modest levels. The viability of the bond market is closely tied to the overall health of the economy, and, although several local governments have recently experienced difficulty in marketing bonds, most feel that such bonds can eventually be sold. In fact, many grantees are more determined than ever to proceed, motivated in large part by the pending reduction in the federal-share ratio. Very recently, considerable interest has risen in using alternative means of seeking private capital for public wastewater facilities. These approaches may eventually offer additional assurance that sufficient capital resources will be available to meet the needs.

The IEPA has attempted to work with local governments to help them prepare to shoulder a greater share of the funding burden. It is encouraging communities to become self-sufficient by planning for adequate reserve capacity (usually a twenty-year growth cycle) and generating enough revenue to maintain and upgrade their facilities. Within the grants programs, for example, this strategy means emphasizing treatment works over new sewer extensions, which are more properly a responsibility of local government or developers. This approach is consistent with the original intent of the Water Pollution Control Act of 1972, which sought not only to catch up with treatment needs but also to encourage communities to plan their capital works programs to satisfy future needs. The 1981 amendments to the act altered this strategy, but the original concept was sound and should be pursued as much as possible. Communities must begin planning and acting now in anticipation of the time when federal and state grant assistance is finally phased down.

Impact of Increasing Local Costs

New investments in wastewater facilities, as well as in the costs of proper operation and maintenance, will mean increased costs for local governments and their citizens. The willingness of people to pay the costs of sewage treatment to meet environmental goals will be a growing issue. Even with the continuation of state and federal grant programs, the local charges to homeowners may be significant.

Local costs may be borne by a relatively small number of people. For some it may be a severe burden. The local share of construction costs is generally financed through bonds, which are repaid using connection fees and user charges or property tax increases. User charges or property tax increases must also finance operation, maintenance, and eventual replacement of the treatment system. The lower the population density of an area, the higher these costs will be per individual home. User charges are a particularly regressive tax: since demand for sewer service is relatively inelastic, user charges have a disproportionate effect on low-income households. As newly constructed systems become operational, increased charges begin. Many communities are now facing such increased costs, and many more will in the near future. Public reaction to the cost of sewage treatment, once it becomes apparent, may have a significant impact on future sewage treatment programs. It is particularly important for the IEPA to assess and communicate the benefits in water quality derived from incurring such sewage treatment costs.

Operation and Maintenance

As significant numbers of wastewater treatment facilities are built, their effective operation becomes an important concern. Construction of modern and complex sewage treatment facilities will only be effective in improving water quality if the facilities are properly operated and maintained. The expenditure of large amounts of federal and state funds on construction grants makes it imperative that the IEPA play an oversight role in the operation and maintenance of sewage treatment facilities to ensure that the public investment is effective in improving water quality. As part of its compliance acitivity, the IEPA will continue to monitor the operation of wastewater treatment facilities, to assist operators informally, and to continue its separate emphasis on operator training and certification programs.

Changes in Regulatory Standards

The need for wastewater facilities represents a set of political decisions on the desirability of specific environmental goals. In the 1970s the public's desire for environmental protection was very strong. Much of this popular sentiment remains but is increasingly tempered by a concern for the costs of abatement and whether or not some goals are realistically obtainable.

In Illinois there is an ongoing effort to reevaluate the state's water quality standards in light of attainable stream uses. The state's current standards are designed to protect: all indigenous aquatic life; such human activities as swimming and fishing; and public water supplies for virtually all waters of the state. However, the waters of the state are diverse in both their nature and their use. Many of Illinois' streams meander slowly through vast, flat agricultural areas, whereas others pass in close proximity to heavily urbanized areas. With this variety of streams, it is obvious that the uses to which they are put will vary to almost the same extent. If streams are to be protected for use as public water supplies, for swimming and fishing, and for industrial use, then stringent water quality standards will be needed. However, many Illinois streams have low or zero flows at most times, and it may not be appropriate to apply water quality standards for public supplies, swimming, or fishing to these streams or stream segments.

The current system of standards does not take this variation into account. Standards are set to protect the most sensitive aquatic species found in the state as well as the most sensitive human uses, regardless of the natural physical and chemical characteristics of a specific stream. As a result, industrial discharges and municipal sewage plants could end up treating their wastes more than is realistically needed to protect the receiving water.

The IEPA believes water quality standards should reflect the practical uses to which a stream may be put. To this end, it initiated a regulatory proceeding before the Pollution Control Board in 1979 to tailor water quality standards to each of the state's 123 stream segments. The IEPA is currently conducting the necessary technical work, including extensive chemical and biological monitoring, to create a new set of standards for each major stream segment.

Discharges from combined sanitary and storm sewers represent the most significant point source water pollution problem remaining in Illinois. The current rule governing combined sewer overflows is a general standard applicable statewide and does not take into account individual situtations in which the nature of the receiving stream might allow a lesser degree of treatment. The IEPA has recommended a modification of the existing rule that would allow it to establish site-specific treatment standards as exceptions to the statewide rule.

Implementation of the Safe Drinking Water Act (SDWA)

Until recently the IEPA's public water supply program has focused on maintenance of the progress begun years ago in protecting public water supplies from waterborne disease. However, implementation of the Safe Drinking Water Act has meant additional stages of program development. The act requires the EPA to establish national water supply standards for contaminants, including those that have not traditionally been monitored or controlled in public water supplies. Of particular concern are organic chemical and radioactive contaminants that pose long-term health impacts rather than short-term dangers from disease or poisoning.

As the SDWA is being implemented, the increasing costs of potable water will become an issue. New standards for drinking water may require costly new water treatment facilities in many communities. Additional laboratory testing and reporting requirements of the SDWA will increase water costs, particularly for small communities. Operation and maintenance costs for water supplies will generally increase and will push water costs up. The federal Clean Water Act's NPDES permit system requirements also apply to water treatment plants with wastewater discharges

and—by requiring wastewater treatment processes not traditionally provided—also add to water costs. These cost impacts vary from community to
community, but many people may experience a large percentage increase in
their water bills.

Water Supply in Northeastern Illinois

Many new communities now have access to Lake Michigan water as a result
of the 1981 amendments to the Supreme Court's decree. Yet even if all these
communities shift from groundwater sources, there will still be a continuing
depletion of the area's deep aquifer. Deeper aquifers than those now utilized have high salinity and would require blending with other water or
extensive treatment. Intensive investigations will have to occur before additional aquifers or surface waters in northeastern Illinois are used as raw
water supplies. Since this area is highly industrialized, there is a greater
potential for chemical contamination of water sources. Sophisticated equipment may be required at raw water intakes both to detect contamination
and to prevent it from reaching the distribution system. The operational
and equipment safeguards necessary in using these less desirable water
sources will result in significant increases in the cost of water to consumers.

Water Resource Management

In Illinois there are nine executive-branch agencies with various responsibilities for water resource management. These are: Environmental Protection Agency, Department of Public Health, Department of Conservation,
Department of Agriculture, Department of Energy and Natural Resources,
Department of Commerce and Community Affairs, Department of Mines
and Minerals, Emergency Services and Disaster Agency, and Department of
Transportation. This fragmentation of water resource programs has in the
past often led to a lack of effective program and policy coordination. In an
attempt to integrate the state's water management efforts and to address
emerging issues, Governor James R. Thompson established the Illinois
State Water Plan Task Force in 1980, directing it to develop "a total water
management system that is socially acceptable and operates within resource
constraints." It is composed of fourteen legislative and executive branch
agencies with water-resource-related programs. The initial focus of this
water planning effort is the identification of significant water issues not
being sufficiently addressed by current programs and emerging issues that
may lead to future problems or conflicts. The task force has identified the
following issues on which to focus its effort:

1. Erosion and sediment control
2. Integration of water quality and quantity management
3. Water conservation
4. Flood damage mitigation
5. Regional competition for water
6. Aquatic and riparian habitat
7. Water-based recreation
8. Drought contingency planning
9. Illinois water-use law
10. Conflict resolution and public participation

The task force has set a schedule of activities for addressing each of these ten issues, which are outlined in a January 1982 progress report to the governor and the general assembly.

Notes

1. James R. Kirk, Jacquelyn Jarboe, Ellis W. Sanderson, Robert T. Sasman, and Carl Lanquist, *Water Withdrawals in Illinois, 1980,* Illinois Department of Energy and Natural Resources, Illinois State Water Survey, Circular 152.

2. See Krishan P. Singh and J. Roger Adams, "Adequacy and Economics of Water Supply in Northeastern Illinois, 1985–2000," Illinois State Water Survey, Urbana, Report of Investigation 97, 1980.

8 Funding Clean Water in Kentucky

Jacqueline A. Swigart and
T. James Fries

In an October 4, 1982, address, Chairman James Howard of the House Committee on Public Works and Transportation described the state of America's public infrastructure as "frightening," and warned that the United States faces "certain economic disaster" unless it establishes a new national policy to rebuild and maintain its multibillion-dollar investment in essential public works. Chairman Howard said:

> It is beyond my comprehension to understand how this administration, or any administration, can hope to revive the national economy, create jobs, control inflation, and maintain a sound defense against aggression, when the basic public works foundation on which all our commerce and industry depend is falling apart.[1]

A plethora of related actions serves to reinforce the attention and emphasis being given to water development financing. First, both the Carter and Reagan administrations have proposed substantial revisions in historically established funding partnership arrangements for water projects; Congress has even supported some of the current administration's proposals, with the most notable being passage of Public Law 97–117, the 1981 Municipal Grant Amendments to the Clean Water Act. Reports such as the Council of State Planning Agencies' *America in Ruins,* The Northwest-Midwest Congressional and Senate Coalitions' *Building a Water Policy Consensus: Key Issues for the Eighties,* and the General Accounting Office's *User Charge Revenues for Wastewater Treatment Plants Are Insufficient to Cover Operation and Maintenance* and *Changes in Federal Water Project Repayment Policies Can Reduce Federal Costs* have contributed to the emergence of water infrastructure financing as one of the key resource issues of the decade.

Problem identification has not been without response. Rather, it has been accompanied by numerous remedial proposals. In addition to legislative proposals S. 621, S. 1095, and H.R. 3432, which were submitted during the first session of the Ninety-seventh Congress, at least four major reconstruction financing bills (H.R. 4366, a bill to create a federal water reconstruction bank; H.R. 7258, the Capital Improvement Act of 1982; H.R. 7265, The Rebuilding of America Act of 1982; and S. 2926, a bill that

109

would also create a National Commission on the Rebuilding of America and develop a ten-year capital investment plan) were filed and referred for committee action during the Ninety-seventh Congress.

The federal government is not the only entity with an interest in infrastructure investment. Meetings such as this Lincoln Institute's roundtable discussion, involvement of the private sector, and the Rockefeller and Donner Foundations' recent award of a $70,000 grant to the Washington-based Clean Water Fund to develop model strategies for financing water and waste management needs in New Jersey, New York, Pennsylvania, and Connecticut; and to examine the role that nonprofit organizations or charitable foundations can play in community financing—more than adequately demonstrate the ubiquitous nature of interest in financial management issues.

Numerous states have reached conclusions similar to those of Congressman Howard. North Carolina and Oklahoma recently joined those states having water financing programs by enacted legislation establishing multimillion-dollar water development funds. Unfortunately, all states have not been successful in their attempts to institute state-level investment programs. For example, the Texas Assembly rejected a legislative proposal to create a water trust fund in the fall of 1981. However, state perception and insight into the need for sound, rational infrastructure development is being demonstrated by the number of water-rich eastern states that have undertaken preparation of water plans not unlike those developed by many of the western states. Kentucky may soon follow in the footsteps of its eastern and western neighbors.

Water Supply and Quality

Until quite recently direct wastewater treatment and water supply infrastructure investment in Kentucky had been a matter left primarily to the local municipal, county, or multicounty project sponsor and the federal funding agency. State involvement was usually limited to occasionally providing technical assistance and reviewing A-95 applications for compliance with state laws and regulations. Three notable noncapital and capital exceptions continue to be: involvement of the state's Public Service Commission, which has responsibility for approving investor-owned utility rates; delegated state administration of the federal Municipal Construction Grant Program; and state financial participation in several Public Law 89–72 recreational cost-sharing contracts.

Largely because of funding and participation changes pending as a result of Sections 3 and 7 of the Municipal Waste-Water Treatment Construction Grant Amendments of 1981, new Farmers Home Administration

loan and grant regulations revisions, and federally initiated budget reces-
sions, deferrals, and reductions, Kentucky has been required to initiate a
reexamination of the state's role in public financing of water project
development, operation, and maintenance. A brief review of two categories
of Kentucky-specific problems and needs should prove illustrative of the
range of financial issues requiring consideration.

Wastewater Treatment

Through FY 1981, between $750 and $800 million had been committed to
the construction of publicly owned wastewater treatment (POTW) facilities
in Kentucky, under the authorities of Public Law 84–660 and the Federal
Water Pollution Control Act Amendments of 1972 (Public Law 92–500, as
amended). By comparison, between $150 and $200 million of Public Law
84–660 funds were spent, while federal and nonfederal Public Law 92–500
commitments totaled some $600 million.

Based on the Environmental Protection Agency's 1978 and 1980 *Needs
Survey* reports, between $460 and $530 million additional would be required
for Kentucky to construct, enlarge, and upgrade the 320 to 400 secondary,
advanced secondary, and advanced wastewater treatment facilities needed
between 1978/1980 and the year 2000.[2] The one-half-billion-dollar figure
does not include any costs associated with sewer rehabilitation, infiltra-
tion/inflow correction, provision of collector and interceptor sewers, and
combined sewer and stormwater control. If these amounts were to be
added, public treatment costs would more than triple. Further, private
treatment system investments are not included in the preceding figures.

The 1981 Clean Water Act amendments covering eligibility and reserve
capacity and the accompanying relaxation of regulatory requirements will
certainly serve to lower treatment cost estimates, but the new estimates will
not be available until the Environmental Protection Agency (EPA) releases
the 1982 *Needs Survey.* Until that time an examination of funding history
and a review of Kentucky's priority list indicates that, as in the past, a high
percentage of Kentucky's annual construction appropriation of approximately $30 million will go to large-flow treatment systems. Further, Ken-
tucky will experience a substantial shortfall in both financial assistance
and treatment facility construction. This is shown by a comparison of a
projected FY 1980 financial need of over $1,100 million, for categories I
(secondary treatment), IIA (advanced secondary treatment) and IIB
(advanced treatment), IIIA (infiltration/inflow), and IVB (new interceptor
sewers), and funding commitments available through the construction grant
phase-out period. This combination of (1) continuing urban industrial bias,
as a result of scale economies, changed eligibilities, priority list direction,

and new nonfederal-share percentages and (2) discontinuance of needed assistance present serious dilemmas that require immediate state attention and action.

Three major investment issues must be dealt with in Kentucky. First, the problem of financing small wastewater system development must be addressed. These systems will experience the greatest problem in raising the newly required up-front planning and design funds and, beginning in FY 1985, the 45 percent nonfederal construction cost share; not surprisingly, these small systems also experience the state's most serious inadequate rate base, debt service retirement, operational cost provision, and replacement fund accumulation problems. Second, a state-level source of funding future project development must be established to supplement federal construction assistance. According to the EPA, thirty-five states had programs to apply state funds to eligible construction grant project costs as of March 1980.[3] Many of the state programs may need to be revised, however, to reflect the requirements of the 1981 amendments. Third, financial incentives and assistance targeted for water quality enhancement through source protection will be needed. This is especially true for non-point source pollution for which small preventive investments are far more cost affordable and effective than expensive after-the-fact treatment.

Kentucky's response to identified wastewater treatment investment needs, although limited at this time, is evolving and expanding. Importantly, an Alternative Financing Work Group has been created within the Kentucky Natural Resources and Environmental Protection Cabinet to examine existing funding sources such as the statutorily established Kentucky Pollution Abatement Authority, and to evaluate alternative financing mechanisms and programs. Further, Kentucky intends to maintain its active involvement through the National Governors' Association and the Association of State and Interstate Water Pollution Control Administrators (ASIWPCA) to positively influence the direction of federal financial deliberations. Last, Kentucky will continue to develop and apply program evaluation and accounting techniques that ensure both financial efficiency and accomplishment.

Water Supply

Many of the same financial problems associated with providing wastewater treatment are also evident with respect to the provision of quantitatively dependable and qualitatively safe supplies of public water in Kentucky. One notable exception is that there is no single water supply counterpart to the federal construction grant program. Rather, there are a variety of smaller-scale financial assistance programs administered by a number of federal agencies. Nonetheless, the problems of system deterioration and needed expansion, economy of scale, inadequate rate structure, and nonexistent state financial assistance require attention if Kentucky is to continue to improve its quality of life and prosper economically. The need for state-level action becomes increasingly significant when there is superimposed on Kentucky's water supply problems: (1) the $75- to $110-billion-investment-requirement figure that was estimated in the Corps' "Nationwide Urban Water System Analysis" study for urban water system improvement from 1980 to 2000,[4] and (2) the General Accounting Office's conclusion that the lack of money is probably the most important factor inhibiting rural water supply development.[5]

One water supply problem recently encountered in Kentucky that is worthy of mention involves an attempt to contract for surplus water in an existing federal flood control and water quality augmentation reservoir project not authorized for water supply. While more than sufficient surplus water is available, new Corp of Engineers (COE) civil works financial policy instruction would require that the water be paid for in a manner that essentially assumes that the project is being constructed at today's cost and interest rates. An additional burden would be the assignment of full project operation and maintenance costs to the water supply contract. Although Kentucky and its communities believe in paying reasonable beneficiary costs, such unrealistic financial demands impede the creation of workable national cost-sharing agreements, interfere with state efforts to ensure sound regional supply development, and impose an inequitable burden on ratepayers.

Kentucky's water supply responses are of two distinct types. First, the Natural Resources and Environmental Protection Cabinet is developing a technical assistance program to aid communities in both demand reduction and supply development. Second, the cabinet's Financing Work Group is analyzing water-supply-funding alternatives and the Kentucky Water Management Task Force, which was created by the 1982 general assembly to examine state water management organization and to assist the cabinet in developing a state water plan, will place a considerable amount of attention on state and local water financing needs, as well as making funding recommendations to the 1984 session of the legislature.

Other Issues

In addition to the water quality maintenance and the municipal, industrial, and rural water supply issues that have been presented, Kentucky has begun to encounter a variety of water-resource-financing problems. Key among these are COE civil works proposals requiring costsharing for planning studies, up-front project financing, potential imposition of new project and program cost-sharing percentages, and discontinuance of funding for existing federal navigation projects. In an effort to affirmatively address these problems, the Commonwealth (1) has established the Kentucky River Task Force, a legislative group charged with determining the appropriate state and federal roles in managing the Kentucky River Commercial Navigation Project, and (2) is chairing the Interstate Conference on Water Problems' (ICWP) Cost-Sharing Task Force.

It can be easily concluded from this review that the commonwealth has accepted its responsibility and initiated deliberate state-level activities to address the issue of financing. It should also be apparent from the identified level of interstate involvement that Kentucky recognizes the need to work cooperatively with all other states to ensure that one-sided federal initiatives do not unfairly or unreasonably burden any state or region.

Funding: History, Problems, and Alternatives

Economic Problems and Needs

Although each state and community experiences unique financial problems, there appear to be four major issues commonly faced in the allocation of public resources among various water investment options and commitments.

The first major problem is affordability. According to the Texas 2000 Commission's report entitled, *Texas Past and Future:* "Capital financing required to meet demands for municipal and industrial water and wastewater treatment during the next quarter century represents an outlay more than double the existing debt of the state and all its political subdivisions."[6] The Texas 2000 conclusion is supported by work of the Texas Water Resource Congress that estimated a state water-financing need of $50 billion over the next twenty-five years.[7] While Kentucky has not attempted to undertake an effort to define the macroeconomic impact of public water investment needs, Kentucky's examples serve to support the magnitude of the Texas conclusions. Without question, all current and projected (inflated) costs for water infrastructure development (planning, administration, operation and maintenance, monitoring, capital construction, and regulatory) must be accurately determined, and conscious decisions and deliberate efforts, including research, to ensure affordability by reducing long-term costs must be pursued.

Second, the problem of regional, collective, and individual equity must be addressed. Although private or individual benefits can be ascertained by market prices, such as increased costs for goods and services and higher taxes, collective benefits from infrastructure investment are not reflected in the marketplace and must be empirically inferred by the amount individuals would be willing to pay. Willingness to pay is, however, highly dependent on each beneficiary's proportionate income and wealth. Additionally, the economic efficiency of social well-being and regional development must be considered in public service investment decisions. All things considered, however, a better balancing of general payments and cost recovery from beneficiaries is absolutely essential. A second collective investment factor that must also be considered is the opportunity cost associated with other competing resource needs of the user area. Another equity aspect to be addressed, as pointed out by General Accounting Office documents and The Northeast-Midwest Institute's report entitled *Building a Water Policy Concensus* is the need for a more equitable distribution of water development funds throughout the nation.[8] The need for improved distribution equity is related not only to geographical areas but also to targeting funds for investment types and purposes. Finally, the question of public equity or financial fairness must also address the appropriate participative responsibilities of all levels of government.

A third major problem involves the soundness of current financial operating practices for public water utilities. The General Acounting Office (GAO), in its 1981 report entitled *User Charge Revenues for Wastewater Treatment Plants* concluded the revenues from user charges were not sufficient to finance operation and maintenance costs and that municipalities are not setting aside sufficient funds to provide for wastewater treatment plant replacement.[9] The GAO also restated its contention from previous plant performance studies that insufficient operating funds have been a major cause of plant operating problems.[10] The report went on to state that Congress agreed to build public treatment plants with the intention that municipalities raise sufficient funds through fair and equitable user-charge systems to operate, maintain, and replace them. In reviewing the legislation initiating the user-charge system, the GAO indicated that "user charges were intended to assure that the financial burden [for wastewater treatment] be spread among all system users in relation to their waste discharge volume and not financed out of local taxes except for *ad valorem* taxes as permitted by the 1977 Clean Water Act Amendments."[11] In other words, Congress intended that wastewater systems be self-supporting. In order to do so, they must generate sufficient revenues to cover annual operations and maintenance costs, pay interest and retire outstanding debt, and make capital outlays required for periodic plant modifications and expansions. To the contrary, the GAO study found that in a majority of cases, the funds needed for plant operation and equipment replacement were coming from revenue sources other than user charges, to include "interest revenue, general funds, and connection, hook-up, and other fees."[12] To correct this serious problem, GAO recommended to Congress that user-charge requirements be made a part of the National Pollutant Discharge Elimination System (NPDES) permit program.[13] Although the solution recommended by the General Accounting Office is not entirely applicable to water supply, several of the capital problems common to publicly financed wastewater systems are also shared by water supply utilities.

A fourth major problem area is the availability of adequate investment resources and financial methods to fund water development. Given current budget constraints and shifts in federal commitment priorities, viable alternative means of raising required capital for water infrastructure development require exploration, testing, and, where appropriate, use at national, state, and local levels.

Alternatives for Financing Future Water Development

Capital-intensive public services such as water supply and wastewater treatment typically display long planning and adjustment periods that may extend over several decades. The consequence of extended lead-time requirements is that government financing is usually insufficient—that is, too little or too much capital is used to adjust to changing population demand and facility duration needs.[14] Congress recognized this problem in the wastewater treatment area when it enacted Section 6 of Public Law 97–117. The section encourages applicants for construction grant assistance to develop capital-financing plans. Further, the EPA has also attempted to address the "planning horizon" problem through development of its "1990 Strategy for Municipal Wastewater Treatment" and establishment of the Financial Management Assistance Program (FMAP).[15] Although efforts to make more effective use of available financial resources are needed, they are not independently sufficient to meet the nation's water infrastructure investment requirements. Even though there are a limited number of financing sources for alternative funding, a review of both current and potential capital and operating revenue sources represents a necessary departure point for any further analysis.

Current Financial Alternatives

Funding and income sources for water supply and wastewater treatment systems can be conveniently identified as: (a) intergovernmental capital, (b) local capital, and (c) local operating.[16] An extensive list of funding alternatives that have been or that can be used to finance water infrastructure is presented by the preceding categories in the following tabulation:

Intergovernmental alternatives
  Conventional federal assistance
    Environmental Protection Agency   Innovative treatment systems grants
                                             Construction grants for municipalities
                                             Loan guarantees for public works con-
                                               struction
    Farmers' Home Administration   Water and waste disposal loans and
                                               grants for rural communities
                                             Rural industrialization assistance
                                             Community facility loans
                                             Industrial development grants
    Housing and Urban Development   Community Development Act Assistance
                                               (including Small Cities Program)
                                             Urban Development Action Grants
    Economic Development Administration   Economic development loans and grants
                                               for public works

  Other federal assistance
    Small Business Administration   Small business water pollution control
                                               loans
                                             Guaranteed pollution control revenue
                                               bonds

  Federal tax incentives

  Regional commissions provision of supple-
    mental loans and grants

  State programs
    Direct subsidies
    Credit assistance
    Grant and loan programs

Local capital alternatives
  Bonds   General obligation bonds
                                             Revenue bonds
                                             Special assessment bonds
                                             Industrial development bonds
                                             Refunding bonds

  Leasing/installment purchase of
    equipment

  Government organizations   Nonprofit corporations
                                             Special districts
                                             Local improvement districts
                                             Regional service authorities
                                             Intergovernmental agreements

  Taxes committed to capital reserve
    funds

  Charges, fees, and permits

  Connection fees   System development fees
                                             Plant investment fees

  Regional revenue distribution

  Private-sector assistance and contributions to
    local government

Local operating alternatives
  Service and users' charges or fees

  General tax revenues   General sales tax

  Special tax assessments   Selective sales tax
                                          Use tax
                                          Ad valorem property tax
                                          General occupation tax
                                          Specific occupation tax
                                          Severance tax
                                          Local income tax
                                          Real estate transfer tax
                                          Site value tax
                                          Land value increment tax

Not all of the listed alternatives are preferable funding mechanisms in all cases, however, and the advantages and limitations of each should be carefully examined before selecting the best alternative to meet specific physical, institutional, and economic needs. For example, connection fees should not be used to pay operating costs when new debt is incurred to build new facilities. Service or user charges, which exemplify the pay-for-service philosophy, represent the preferred method for providing local operating costs because they can be set to reflect the full cost of operation, the payment of debt service, and the accumulation of depreciation funds necessary for plant and equipment replacement. Reliance on user charges also avoids the need to subsidize operating costs from general-fund tax revenue and ensures that limited sources of local funds for general-purpose, non-revenue-producing activities are not diverted. Although revenue-generating programs that institutionalize the flow of funds through small surcharges on water and wastewater services are difficult to develop and more difficult still to have accepted by the public, they appear to be one of the best long-term solutions for ensuring continuing financial support and facility viability.

Another key aspect in financial mechanism selection is ability to pay, which in turn is closely linked to the issues of affordability and equity. Evaluation of any alternative must take into consideration both the sponsoring agency's and the individual system user's ability to pay the amount necessary to provide the service (cost). Factors to be considered in an analysis of financial capability include sources and amounts of revenue and expenditure, debt history, tax rates, user fees or charges, and the needs of competing programs.

An extremely important financial tool related to ability to pay that must continue to be examined is pricing or rate structure. Alternative rate strategies deserving increased attention include value-of-service pricing and social pricing. When coupled with traditional cost-of-service pricing, both appear to offer needed and desirable features. Finally, the use of inverted (increasing) block rates and peak-load rates need to be considered as alternatives to the more commonly used uniform and decreasing-block-rate-pricing approaches.

Alternative Financial Arrangements

The range of restructured and new financial alternatives appears to be more limited in most cases than the approaches now available and in use. At the national level, proposals have been made to develop a national capital budget and to establish a water utility bank and a federal reconstruction bank that would fund long-term water development. Congress is currently considering proposals to meet urban water needs through tax credits and leasing arrangements. Aid for rural water systems through continuing existing Farmers' Home Administration, Housing and Urban Development, Economic Development Administration, and other regional development programs and establishing new low-interest loans are also being considered actively.

At the local level the increased use of enterprise bonds, the development of sale-and-lease-back transactions, the provision of nonprofit corporation and vendor-assisted financing, and parcel packaging are all receiving popular attention. Also, private-sector investor financing of public treatment works (privatization) appears to offer distinct advantages in certain community situations.

In addition to examining specific public investment approaches and mechanisms, any discussion of wastewater financing must examine the different forms of environmental pricing that can be used to address economic disincentives. Four approaches are recognized as available for use.[17] First and most frequently used in the United States is direct regulation. While regulation may offer great certainty in achieving specified levels of treatment, its cost is high, and the regulated community is encouraged to postpone control or abatement decisions as long as possible. A second approach involves providing incentives for controlling discharges. Effluent charges, consisting of both user or service charges and surcharges, comprise a third control approach. Fourth, punitive and excise taxes can be employed as a disincentive to continuing uncontrolled discharges. Although it is unlikely that federal and state emphasis will be shifted away from direct regulation in the near terms, the General Accounting Office reported in 1978 that fee systems provide a continuing incentive to control discharges and can reduce control costs and increase administrative and enforcement effectiveness.[18]

Related Control Alternatives

The provision of a sound water infrastructure base is not exclusively an economic issue. It also involves important considerations relating to supportive institutional arrrangements and controls. For example, special districts and associations are created to provide public services on an efficient scale and to capitalize on size economies. Subdivision regulations are frequently used to impose infrastructure cost burdens on developers. Agreements between utility operating districts and land management agencies are formulated on the basis of future service areas, and cities and counties exercise their land-use regulatory powers (comprehensive plan, zoning ordinances) to ensure that growth patterns support past and planned investment in utility services.

Of the alternatives available, legislative adoption of utility system extension policies appears to be the single most effective form of institutional control because such policies can be used to prioritize and direct facility expansions and to establish the conditions for adding new customers to utility systems. Extension policies can be both passive and active. A *passive policy* identifies an efficient service area and denies service provision elsewhere; *active policies* provide directional incentives by building facilities and providing service to desirable expansion areas. The use of active policies requires, however, that communities be committed to eliminating reactive decision making and concentrate instead on anticipation and advance design and implementation of rationally selected, desirable directions and actions. Without question, the political costs of such a philosophical orientation are high.

Another form of useful institutional control involves the negotiation of advance service contracts between utility systems and new service parties. Important financial aspects of contractual service arrangements negotiated under an approved utility plan include developer financing, developer payback or reimbursement provisions, and assured occupant or improvement district repayment and service charge funding.

Added advantages associated with all the institutional alternatives reviewed include the provision of a clear responsibility for system management, and the separation of publicly operated utility systems from general government activities. The second aspect promotes economic self-sufficiency of the utility and discourages the diversion of utility income and revenue to support programs normally funded by general taxation.

Conclusions

Although the exact extent, severity, and cost of national, state, and local infrastructure deterioration has yet to be determined, the problem promises to profoundly affect the nation's and nearly every citizen's future for decades to come. Characteristic of the past, individual state government actions and responses indicate a strong commitment to addressing the problem and finding workable solutions, while federal government action, due in large part to competing national priorities of defense and development, continues to be fragmented and delayed. The failure to reach a common commitment and to agree in advance on universally desirable approaches serves only to penalize entities pursuing progressive efforts to resolve the problem.

Because of the tremendous potential cost associated with public water infrastructure rehabilitation and provision, every proposed solution must address the questions of affordability, equity, future financial self-sufficiency, and capital availability. Lack of consideration for any one factor may jeopardize or impair the potential for success.

At the same time the future of financing does not depend exclusively on designing new approaches. Rather, it requires both employing existing financial mechanisms and practices and developing creative economic and institutional alternatives. A better and more aggressive sharing of information—such as the opportunity provided through the Office of Management and Budget's Federal Assistance Program Retrieval System (FAPRS)[19]—and experience between all states and units of local government is an essential, constructive component of the financial learning process.

Notes

1. House Committee on Public Works and Transportation, *News From,* October 4, 1982, p. 1.

2. U.S. Environmental Protection Agency, *1978 Needs Survey,* pp. 47–81 and p. 23.

3. U.S. Environmental Protection Agency, Office of Water Program Operations, *Clean Water Fact Sheet,* July 1981, p. 5.

4. U.S. Army, Corps of Engineers, *Urban Water Systems: Problems and Alternative Approaches to Solutions,* submitted by the Subcommittee on Urban Water Supply of the President's Intergovernmental Water Policy Task Force, p. II-11.

5. U.S. General Accounting Office, *Rural Water Problems: An Overview* (CED-80-120), p. 20.

6. *Texas Past and Future: A Survey,* quoted in Information Water Service, Inc., *Water Information News Service,* September 20, 1982.

7. Texas Water Resources Congress, quoted in Information Water Service, Inc., *Water Information News Service,* September 20, 1982.

8. Northeast-Midwest Institute, *Building a Water Policy Consensus:*

Key Issues for the Eighties (Washington, D.C.: Northeast-Midwest Institute, 1982), pp. 5–21.

9. U.S. General Accounting Office, *User Charge Revenues for Wastewater Treatment Plants—Insufficient to Cover Operation and Maintenance* (CED-82-1), p. 8.

10. Ibid.

11. Ibid.

12. Ibid.

13. Ibid., pp. iii–iv.

14. Stephen P. Coelen, "Public Service Delivering in Rural Places," *Rural Development Perspectives,* September 1981, p. 23.

15. U.S. Environmental Protection Agency, Water Planning Division, *Planning for Clean Water Programs: The Role of Financial Analysis,* p. 4 (U.S. Government Printing Office, 1981).

16. Ibid., p. 15.

17. Senate Committee on Environment and Public Works, *Pollution Taxes, Effluent Charges, and Other Alternatives for Pollution Control,* May 1977, pp. 10–11.

18. U.S. General Accounting Office, *16 Air and Water Pollution Issues Facing the Nation* (CED-78-148B), p. 100.

19. U.S. Environmental Protection Agency, *Planning for Clean Water Programs,* p. 16.

9

Funding Clean Water in Maryland

Mavis Mann Reeves

Water Supply and Quality

Water in Maryland is relatively plentiful—indeed, water covers almost one-fifth of the state. Moreover, even though the state is rather densely populated, Maryland's water supply problems are less severe than those of many states since industrial and irrigation demands have been comparatively low in recent years because of a general decline in population growth in many areas of the state. This situation is likely to change very soon, however, for a 20 percent increase in demand is anticipated by 1990.[1]

The vast majority, about 90 percent, of Maryland's water supply comes from surface water. The Piedmont and Appalachian regions have excellent surface water and reservoir sites. The Coastal Plain area, where surface water and impoundment sites are scarce, is underlain with plentiful aquifers.[2] A major exception to sufficient water is in Charles and Saint Marys Counties in southern Maryland. There, present supplies are adequate and apparently will be through 1990; however, groundwater withdrawals may exceed recharge rates by the turn of the century, or saltwater intrusion may make the water unusable.[3]

Pollution and wastewater treatment are more serious problems. Because of the state's small size and geographic location, its rivers and the Chesapeake Bay fall prey to pollutants from both inside and outside its boundaries. Chesapeake Bay, in particular, is threatened seriously by pollution from accumulated toxic substances and chemicals from agricultural fertilizers and from other non-point sources of pollution such as the runoff from city streets. In the realm of wastewater treatment, many of Maryland's facilities are old and may soon be in need of replacement.

Water Management and Infrastructure

Necessity for Intergovernmental Cooperation

In Maryland's efforts to secure and maintain clean water, cooperation with other governments is a necessity. Because of its geographic situation, the

The author acknowledges the assistance of Kenneth McElroy, Jr., Office of Environmental Programs, Maryland Department of Health and Mental Hygiene, and Professor Stephen W. Sawyer, Department of Geography, University of Maryland, College Park.

state is heavily dependent on the actions of officials in Virginia, West Virginia, Pennsylvania, the District of Columbia, the federal government and, to a lesser degree, New York, in its water supply and pollution control activities. Maryland officials deal with several national agencies in addition to the ones from whom they might seek financial aid in confronting the state's water problems. The Army Corps of Engineers plays a major role. In supplying water to Washington, D.C., the Corps integrates its plans with those of the adjoining jurisdictions, including several in Maryland, dependent on the Potomac River. The Corps also assists the state with planning water resources under Section 22 of the Water Resources Development Act of 1974.[4]

The Interstate Commission on the Potomac River Basin

Maryland participates in two interstate commissions affecting water supply. The Interstate Commission on the Potomac River Basin was established in 1940 to address water quality problems in the area. Essentially an advisory and research organization, the commission monitors water quality in the basin, points up critical issues, engenders cooperation among area jurisdictions, and advises on water supply conditions for the Washington, D.C., metropolitan area. Attempts to adopt a stronger interstate program for regulating water activities in the basin failed when West Virginia and Pennsylvania, at the headwaters of the Potomac, refused to ratify the Potomac River Basin Compact.[5]

The Susquehanna River Basin Commission

The Susquehanna River Basin Commission incorporates less of Maryland's area but exerts stronger control in the state. It can regulate to secure the adoption and implementation of its decisions. The commission's requirement that during low-flow conditions all water withdrawers must return to the river as much water as they use has improved the reliability of supplies of municipal systems dependent on the Susquehannah and has helped to stabilize the ecology of the upper Chesapeake Bay.[6]

Precompact withdrawal rights of Baltimore and approximately twelve other large jurisdictions were recognized by the commission, although it can suspend all such agreements in the event of an emergency.[7]

State Agency Involvement

Several state agencies have major responsibilities in both water supply and wastewater treatment. The Office of Environmental Programs (OEP) in the Department of Health and Mental Hygiene is heavily involved. It enforces regulations of the U.S. Environmental Protection Agency (EPA) relating to water supply systems covered by the Safe Drinking Water Act of 1974 (those that serve more than twenty-four people or have more than fourteen connections). The OEP has been designated as the regional agency responsible (under Section 208 of the Federal Clean Water Act of 1972) for coordinating wastewater treatment planning for all areas of the state outside the Washington and Baltimore regions as well as the state agency responsible for statewide water quality management. It also closely monitors well and surface water quality levels, certifies the water in new well sites for human consumption, licenses all well and filtration plant designs, investigates potential threats to water supplies such as landfill sites, evaluates county water and sewer plans, and is the lead agency in any drought or equipment outage that threatens public health or safety such as fire protection. The OEP has the authority to shut down any supply source or to require changes in its operation, and the secretary of the Department of Health and Mental Hygiene can withhold permits for water facility construction for counties without approved water and sewer plans.

In regard to wastewater treatment, the OEP may order a municipality or private industry to construct a new sewage treatment plant or to rehabilitate one in operation. It also develops and manages the state priority list of wastewater treatment projects eligible to receive federal funds. In this case the locality may propose a project and the OEP decides its placement on the list.[8]

Water supply planning is one of the central functions of the Water Resources Administration in the Department of Natural Resources. It is concerned with erosion control, wetlands protection, dam safety, floodplain hazard management, dredging, hydraulic impacts of mining, and facility supply and water supply issues. Its Water Appropriation Permit Section is concerned with the water resource base and any conflicts between water withdrawals and existing water uses. In this connection its activities are primarily regulatory.

The Maryland Environmental Services, also in the Department of Natural Resources, is a quasi-public corporation that operates as a public utility company for the state. It designs, plans, finances, constructs, and operates facilities for wastewater treatment, water treatment, and solid waste disposal (and certain other facilities at state institutions). In addition, it may temporarily run some private or municipal facilities that have not been able to provide adequate service. Currently it operates twenty-six water supply facilities, seventeen of which are state owned. Nonstate facilities were taken over at a county's request usually because the original owners were not operating them competently. Typically, these systems are very small, aging, and deteriorating or deficient in design.

The Maryland Geological Survey, also in the Department of Natural

Resources, conducts geologic and hydrogeologic investigations for the state, prepares maps, and disseminates information on water resources. It is particularly concerned with "identifying changes in present supplies [river and groundwater monitoring], potential new sources of supply, yields of different aquifers, and impacts of excessive ground water withdrawals, and future conditions."[9]

The Maryland Potomac Water Authority was established to finance the nonfederal share ($8 million) of the construction costs of the Bloomington Dam Reservoir on the North Branch of the Potomac River. It has authority to sell water stored in the dam.

Although not directly concerned with water and wastewater treatment programs, the Department of State Planning engages in activities that directly relate to clean water. This agency has primary responsibility for advising the governor on all planning matters in the state, including those relating to major development activities and on the location of state facilities. Of particular interest in a discussion of clean water are the department's responsibilities for preparing the state's capital budget, reviewing county water and sewer plans for conformance with local master plans, and projecting state development needs and potential natural resource management issues. In the past it also has served as the clearinghouse for federal grant applications.

Substate Handling of Water Matters

Compared to most states, Maryland has relatively few local governments. It ranks eighth from the bottom in total number—a situation that helps to mitigate coordination problems. Moreover, the state places local planning responsibility relating to water supply and sewerage in the hands of the twenty-three counties and Baltimore City (distinct from Baltimore County, although it is considered a county under Maryland law), which reduces still further the jurisdictions with which the state must deal in these matters.

Nevertheless, substate water management in Maryland is far from uniform. A report prepared for the Maryland Commission on the Functions of Government in 1973 stated that "the one uniform statement that can be made is that very little uniformity exists among the variety of systems. In both water supply and sewage disposal, each local jurisdiction appears to chart its individual course in areas such as physical facilities and methods of treatment, organizational arrangement within the local government structure, rate-setting, means of raising revenues to finance both capital and operational costs, and relationships."[10]

Approximately 625 community water systems and thousands of individual wells provide Maryland citizens with water. The two largest of these systems—those of the city of Baltimore, which supply the surrounding area, and the Washington Suburban Sanitary Commission (WSSC), serving Montgomery and Prince George's counties on the fringes of Washington, D.C.—serve 70 percent of the state's population. According to a report prepared for the Department of State Planning in 1981, both of these suppliers have "long histories of excellent management and far-sighted planning," and consequently are not burdened by serious system deficiencies and planning errors. The report also noted that municipal systems, in general, are run by skilled professionals.

The WSSC and the city of Baltimore's Department of Public Works handle sewage matters for their respective regions. In other jurisdictions, water and sewer matters typically are handled by a department of public works.

Regional planning agencies play important roles in substate water and sewer management. The Regional Planning Council serves Baltimore City and Anne Arundel, Baltimore, Carroll, Harford, and Howard Counties. The Washington Metropolitan Council of Governments draws membership from Virginia and the District of Columbia as well as Maryland jurisdictions in the region. Most of the activities of these two agencies relate to planning and encouraging cooperation within their respective areas. The Regional Planning Council emphasizes water quality and is especially concerned about non-point source water pollution. It also intervenes in disputes between jurisdictions that are served by the Baltimore water system and those that are not. The Washington COG coordinates water supply activities among the sixteen jurisdictions supplying water in its region, and it played an important role in the adoption of the Water Supply Emergency Agreement for regional cooperation during droughts or equipment failures. Other regional planning councils also are active.

A Strong Planning Tradition

Maryland's strong planning tradition has contributed to better management of water and sewer facilities. Its Department of State Planning, created during the 1930s with funds from the National Planning Board, continued to function when most other states abolished their planning agencies. It still exercises strong influence in state policy matters because of a close relationship with the governor's office.

Since 1966 Maryland has required every county to adopt and submit to the Department of Health and Mental Hygiene a ten-year water and sewer plan listing water and sewer facilities, their costs, and their relationship to local land use and master plans. Other local water and sewer systems are included in this plan, and future installations of community water supply systems, sewerage, or solid waste disposal or multiuse water and sewage disposal systems must be in accordance with the plan.[11] Every county has had a plan since 1970.[12] The state is able to enforce this provision by withholding construction permits for projects that do not meet its requirements and by prohibiting subdivision and building approvals that overtax the capacity of existing facilities.

The federal Water Pollution Control Act of 1972, which provided funds for planning as well as for construction, also contributed to good planning. An official at the Department of State Planning is reported to have commented that "detailed planning and planning coordination was carried out, and with beneficial effects in many, if not all, cases. . . . There were many examples where interjurisdictional coordination replaced years, even decades of bickering."[13]

Nevertheless, water planning in Maryland has been criticized. In a report yet to be released, the Inspector General of the U.S. Environmental Protection Agency reportedly called the sewage treatment program in Washington, D.C.—and its Maryland suburbs—a failure. According to a *Washington Post* account, local officials failed to "plan, design, and construct needed facilities." Allegedly, the Inspector General advocated that the entire effort be turned over to the federal government, pointedly criticized the Washington Suburban Sanitary Commission, which serves Montgomery and Prince George's counties, Maryland, and bemoaned political pressures exerted in the Maryland suburbs that delayed or prevented the construction or operation of proposed sewage treatment and sludge disposal plants. This assessment was disputed by a WSSC official who pointed to the clean-up of the Potomac River—where swimming and fishing are now possible—as an indication of success.[14]

According to a recent consultant's report, a major problem for infrastructure planning in Maryland is the lack of adequate information. Although data on water and sewer facility needs were more often available than those in other functional areas, the report stated, officials of small communities often do not have the information to make cost-effective decisions on water and sewer facility investments. Almost half of those surveyed believed they could establish cost-effective priorities for water supply facilities, and one-third believed they could do so for sewers. Information on storm sewers was regarded as the least reliable of all water and sewer needs. Not surprisingly, more metropolitan area respondents rated their information higher than did those from nonmetropolitan areas.[15]

Although Maryland has a water conservation law, enacted in 1979,[16] requiring the installation of water-conserving equipment in buildings and prohibiting the sale of plumbing equipment that is not water conserving, a majority of respondents to a 1981 survey of water supply system managers expressed "no need for or interest in water conservation programs." This judgment applied to those systems having public education programs as well as those imposing higher rates for increased usage.[17] They gave several reasons for their attitudes. Many smaller systems generally have excess water because they have had to construct their systems to meet relatively large fire-fighting needs. In others, managers welcome the increased funds that higher consumption provides and fear that increased rates would not compensate for lower consumption. Slow-growing communities want to hold and attract industries. Lastly, many believed that higher rates would be politically unpopular.

Perceptions about their water supply situations also contributed to the respondent's lack of interest in conservation. When questioned about the possibility of cutbacks if a repeat of the 1966 drought occurred, twenty-three of the thirty-five respondents replied that none would be needed, four thought modest voluntary reductions might be required, three reported that both voluntary and "modest" mandatory cutbacks would be necessary, and five said they would anticipate major use reductions and extensive mandatory restrictions. The last-named group of respondents included the WSSC, where construction of the projected Seneca Creek reservoir should boost water availability in the area. It should be noted that WSSC has one of the two most active conservation programs in the state; the other is in Howard County, which employs a full-time conservation specialist. Several other systems have conservation programs, and several systems impose rates that penalize heavy use.

A deterrent to public efforts at conservation education is the nonparticipation of some large industrial users that have inhouse programs aimed at cutting water costs. Both Bethlehem Steel in Baltimore and Bauch and Lomb in Oakland, for example, now rely on renovated or recycled water for most of their needs. The renovated water used by Bethlehem Steel equals 52 percent of the Baltimore system's consumption. An overwhelming majority of respondents (thirty-three) indicated that water users in their areas would not be receptive to the use of recycled water.

Water Facility Needs

Little doubt is expressed about the ability of current water supply systems to meet future needs. A 1981 survey of representatives of the thirty-five water supply systems that serve more than 4,000 people—and supply 87 percent of the state's population—found them "universally confident" of their ability to meet expected demands for 1990.[18] As previously noted, they anticipated an overall growth of 20 percent in water demand. An increase in usage of less than 5 percent was expected in six systems, while seven communities looked for growth of more than 80 percent. Although the largest rises in demand were forecast for Anne Arundel County, the Baltimore system, and the Washington Suburban Sanitary Commission, Charles and Saint Marys counties were found to face the most serious supply problems. There, groundwater aquifers may not be adequate to meet demands if the populations of the counties continue to grow at their present rates. Large-scale impoundment is prohibited by the topography of the area, and a suitable alternative has not yet been found.

In the low-growth regions of the state—western Maryland and the Eastern Shore—present capacities can meet projected needs. Other areas plan to rely on the expansion of supply sources, largely by drilling wells or developing additional storage or impoundment capacity. Frederick County is the only system anticipating seeking additional water allocation.

This optimistic view of supply is reenforced by opinions of five non-state agencies, also surveyed in 1981. The consensus was that no imminent supply problems exist. Minor supply problems, as well as the most serious planning weaknesses, are thought to exist among the small community systems.[19] No statewide estimates of water supply facility needs are available at this time. It is interesting to note, however, that only four system respondents cited the need for state assistance.

Wastewater Treatment Facility Construction Needs

The U.S. Environmental Protection Agency and the state of Maryland reached identical conclusions when they estimated the construction costs of publicly owned wastewater treatment facilities required to serve the population of the state as it is today and as it is projected to be in the year 2000.[20]

Costs of construction needed in 1982, sometimes referred to as backlog needs, were estimated at $830 million, or $179 per capita. This figure does not include the $2.9 billion required for controlling urban stormwater runoff, which would add another $691 per capita (for a total of $870 for each resident). The Maryland costs, not including the stormwater runoff needs, amounted to 0.89 percent of the total national needs, and Maryland ranked twenty-eighth in the nation (see table 9-1 for the estimates).

By the year 2000 a total of $1.5 billion—or an additional $206 per capita—will be needed for construction costs for wastewater treatment facilities, correction of infiltration or inflow, major rehabilitation of sewers, and new collector and interceptor sewers. This amounts to a total of $897 per capita. Because the needs are not spread evenly through the state, per capita costs in some communities will be substantially heavier than in others if the projects are locally financed. No estimates for the year 2000 are available for control of combined sewer overflow or for stormwater runoff.

Table 9-1
1982 EPA/Maryland Estimates of Construction Costs of Publicly Owned Wastewater Treatment Facilities
(millions of dollars)

	Total (I–V)	I	IIA	IIB	IIIA	IIIB	IVA	IVB	V	VI
Backlog Needs[a]										
1982 estimates	$830	$106	$227	$0	$13	$130	$225	$51	$75	2,017
Change from 1980	−2,093	−146	−188.3		+1.15	−1,715	9.3	−30.3	−23.0	−363
Percent of national total	0.89	0.53	6.03		0.54	2.79	1.35	0.57	0.21	3.13
Year 2000 Needs[b]										
1982 estimates	$1,152	$194	$246	$17	$13	$130	$266	$282	n.a.	n.a.
Change from 1980	0	−186	−177		+1.15	−1715	+7.9	−141	n.a.	n.a.
Percent of national total	1.037	0.62	4.06		0.53	2.79	1.29	1.59	n.a.	n.a.

Categories

I. *Secondary Treatment:* Costs for facilities to achieve secondary levels of treatment, regardless of the treatment levels required at the facility site and costs for systems designed to serve individual residences.

IIA. *Advanced Secondary Treatment* (AST): Incremental costs above secondary treatment levels to achieve advanced secondary levels of treatment for those facilities that must achieve such levels. This requirement generally exists where water quality standards require removal of standard pollutants at higher levels than 85 percent or 30/30; but less than 95 percent removal or 10/10.

Table 9-1 (continued)
(millions of dollars)

IIB.   *Advanced Waste Treatment:* Incremental costs above AST for those facilities that require advanced levels of treatment. This requirement generally exists where water quality standards require removal of such pollutants as phosphorus, ammonia, nitrates, or organic and other substances or where removal requirements for conventional pollutants exceeds 95 percent.

IIIA.   *Correction of Infiltration/Inflow:* Costs for correction of sewer system infiltration/inflow problems. Costs could also be reported for a preliminary sewer system analysis and for a detailed sewer system evaluation survey.

IIIB.   *Major Rehabilitation of Sewers:* Replacement and/or major rehabilitation of existing sewer systems necessary to the total integrity of the system. Major rehabilitation is considered to be extensive repair of existing sewers beyond the scope of normal maintenance programs where sewers are collapsing or structurally unsound.

IVA.   *New Collector Sewers:* New collector sewer systems and appurtenances designed to collect violations caused by raw discharges, seepage to waters from septic tanks, and/or to comply with federal, state, of local actions.

IVB.   *New Interceptor Sewers.* New interceptor sewers and pumping stations necessary for the bulk treatment of wastewaters as well as outfall sewers.

V.   *Control of Combined Sewer Overflow:* Facilities to prevent and/or control periodic bypassing of untreated waste from combined sewers.

VI.   *Control of Urban Stormwater Runoff:* Cost estimates for treating or controlling pollution from separate storm sewers. No federal grants are made for this category.

Source: Compiled from U.S. Environmetnal Protection Agency, Office of Water Programs, *1982 Needs Survey: Cost Estimates for Construction of Publicly-Owned Wastewater Treatment Facilities.*

n.a. = not available.

[a]Costs for backlog needs are sufficient only to build facilities for 1980 population and exclude reserve capacity for future population growth.

[b]Costs for year 2000 needs include cost of construction required to meet population projections at that date. The population for Maryland is projected to be 5,583,000.

As table 9-1 indicates, construction cost estimates dropped considerably from 1980 to 1982. Backlog needs, excluding stormwater runoff, reflected in the 1980 figure, were estimated at $3.0 billion compared to $830 million in 1982. The largest reductions relate to major sewer rehabilitation—shown as a major need in the 1980 estimates—and facilities for wastewater treatment. At the same time estimates for correction of infiltration rose during the biennium, although the increase was small.

Wastewater treatment facility funds and new collector sewers constitute the most costly needs in categories for which federal grants-in-aid are available. Their estimates are dwarfed, however, by the $2.5 billion required for control of urban stormwater runoff.

Funding: History, Problems, and Alternatives

Funds for water supply and wastewater treatment systems in Maryland currently come from several sources. Some construction costs are met with user fees—appearing on the water and sewer bills of consumers—and local governments in Maryland are empowered to issue bonds for water and sewer facilities. Much of the financial backing comes from intergovernmental funds, and both the state and the federal government contribute to construction through grants and loans.

With the possible exception of Charles and Saint Marys counties, local systems appear to be equipped to handle part or all of the construction costs for water supply. In the 1981 survey noted earlier, most system representatives anticipated no major difficulties. Eight cited potential problems connected with system expansion, but for the most part, they expect to raise rates to pay for the changes. Their optimism is encouraged by low average water bills in Maryland: for a family of three, average annual rates run to $104, leaving room for additional rate raising. Across Maryland, annual water bills range from $40 in Centreville on the Eastern Shore to $230 in Oakland in the extreme western part of the state.[21] It should be noted, however, that rates reflect no particular geographic pattern. Although many of the suppliers that charge the highest rates are in the North and Northeast, the higher rates are scattered throughout the state. Higher rates have been imposed in about one-third of the systems in recent years.[22] In fact, some of the expansion and reconstruction that has occurred in recent years has been financed, at least in part, in this manner.[23]

Three towns—Oakland, Brunswick, and Prince Frederick—expressed concern over being able to handle future demands without an improvement in their financial status. Consultants for the Department of State Planning, however, have asserted that these problems are not particularly serious because current supply sources are plentiful and only minimal growth is

anticipated. The older water distribution systems in Baltimore City, Brunswick, Hagerstown, and Oakland are deteriorating, but currently this is not
regarded as a major problem.

Financing wastewater treatment facility needs is another matter since
localities will have difficulty handling these costs without assistance. Estimated backlog costs for facilities needed now run $830 million, and costs
for the year 2000 are projected at $1.1 billion. This amounts to $179 per
capita for backlog costs and approximately $220 per capita by 2000. The
needs are not distributed evenly, however, and the costs in some areas will
be much higher. (It should be noted that these figures cover only publicly
owned wastewater treatment facilities and do not include private systems.
Neither do they include costs of controlling urban stormwater runoff, estimated at $2,917,109 in backlog needs. No federal funds are presently
available for these activities.)

The funding mix in the past has been 75 percent federal (to decline to 55
percent in the next few years), 12.5 percent state, and 12.5 percent local.
The federal share was increased to 85 percent in 1978 for projects that were
deemed innovative.[24] In addition, Maryland will contribute 11.25 percent of
eligible costs for alternative innovative projects, leaving local governments
with innovative systems to pay only 3.25 percent of facility costs.[25] (These
figures apply only to wastewater treatment programs handled by the EPA;
other available federal funds, such as grants or loans from the Farmers'
Home Administration for construction of community facilities, are not
included.

The extent of future federal funding is uncertain. To date, Congress has
rejected proposed cuts in EPA funds for water pollution control grants.
How they will fare in the current economic climate—when they have to
compete with human needs for welfare and health care and the defense
build-up—is, of course, unknown.

The state has established a Sanitary Facilities Fund that provides loans
for the difference between total costs of a project and the total of all federal
funds applied to it. These funds come from specifically authorized sanitary
facilities sewerage bonds.[26] The 1982 legislature reallocated monies in
another state fund, the Water Quality Loan Fund, established in 1974, to
increase from $5 million to $10 million the money available for state loans
for sewer construction facilities. Localities can receive grants, loans, or loan
guarantees from this source for the acquisition, construction, equipping,
rehabilitation, and improvement of sewage treatment plants and related
facilities, water supply facilities, solid waste disposal, and agriculturally
related non-point pollution control projects.

The state continues to recognize its responsibility for clean water.
Governor Hughes directed the State Development Council, which is com-

posed of certain cabinet-level officials, to study the infrastructure situation in Maryland and to make recommendations that will include possible legislative actions. The council has established a committee to identify capital needs in several areas, including water supply, wastewater, stormwater, sediment and erosion control, and non-point source pollution control.

If the state decides to assist local governments with water and sewer projects to a greater extent than it now does, it is in a better position than most states to do so. Although Maryland, like other states, has felt the pinch of the recession and the burden of increased needs and has had to work to balance its books, it has both lower unemployment and higher income levels than the national averages. Using the Representative Tax System developed by the U.S. Advisory Commission on Intergovernmental Relations, however, its tax capacity is just about average while its tax effort is about 9 percent above the average for all states.[27] The state's bond rating is sufficient to permit further borrowing at the lowest municipal rates.

Looking to the Future

Despite the somewhat favorable state outlook the most severe problem Maryland will face in connection with providing clean water is the financial one. In some instances, it may be necessary to choose whether to moderate standards or to increase outlays. The prospect of diminished federal assistance will serve as a deterrent to a number of projects, particularly in small communities, and the pressure on the state to increase financial support is likely to rise. Greater interlocal cooperation and improved cost data will be necessary.

Although the outlook for improved water and sewer facilities in Maryland is far from being problem free, there is some basis for optimism. Both state and local officials are aware of the importance of infrastructure maintenance, replacement, and expansion, and the state is already engaged in assessing the needs and costs and in upgrading planning in this connection. Moreover, current state law requires municipal governments to upgrade deficient, publicly owned sewage treatment systems, something that has engendered little resistance on the part of local officials. As one state environmental officer pointed out, "No Maryland town or city has been hauled into court over its failure to upgrade properly its effluent discharge system as required by state law."[28]

In most instances state prodding appears unnecessary. A 1980 survey revealed that municipal officials accorded water supply and sewage treatment the highest priority among sixteen listed functions.[29] Moreover, they have often sought financing on their own.

Notes

1. Donald E. Petzold and Stephen W. Sawyer, *The Structure and Status of Water Supply Planning in Maryland* (Baltimore: Maryland Department of State Planning, July 1981), p. 15.

2. Ibid., pp. 53–54.

3. Ibid., p. 16.

4. Ibid., pp. 36–38. On the role of the Corps in the Potomac River Basin, see: Senate Committee on Governmental Affairs, *Potomac River Basin. Regional Planning: The Potomac Experience,* Committee Print, 97th Cong. 1st sess., September 1981. Prepared by David K. Hartley, Governmental Affairs Consultant, Congressional Research Service.

5. James A. Mederios, "The Politics of Water Resources Development: The Potomac Experience" (Ph.D. diss. University of Maryland, College Park, 1969), pp. 113ff.

6. Petzold and Sawyer, *The Structure and Status of Water Supply Planning,* p. 43.

7. Ibid.

8. Brian M. Gardner, "Intergovernmental Fiscal Relations and Local Government Policy: A Study of Wastewater Treatment Grants in Maryland" (Ph.D. diss. University of Maryland, College Park, 1981).

9. Ibid., pp. 185–186.

10. Maryland Commission on the Functions of Government, *Information Report on Water and Sewerages* (Mimeographed) (Annapolis, Md., October 19, 1973), pp. 11–12.

11. Petzold and Sawyer, *Structure and Status of Water Supply Planning,* p. 64.

12. Garfield and Schwartz Associates, Inc., *Local Infrastructure Planning in Maryland* (Baltimore, Md.: Maryland Department of State Planning, 1982), p. 32.

13. Ibid., p. 33.

14. "EPA says D.C., Maryland Sewage Efforts a Failure," *Washington Post,* January 21, 1983, p. B5.

15. Garfield Schwartz Associates, Inc., *Local Infrastructure Planning in Maryland.*

16. *Annotated Code of Maryland,* Art. 43, Sec. 325D.

17. Petzold and Sawyer, *Structure and Status of Water Supply Planning,* pp. 46–48.

18. Ibid., p. 1.

19. Ibid., p. iii.

20. U.S. Environmental Protection Agency, Office of Water Program Operations, *1982 Needs Survey: Cost Estimates for Construction of Publicly-Owned Wastewater Treatment Facilities,* December 1982.

21. Petzold and Sawyer, *Structure and Status of Water Supply Planning,* p. 20.

22. Ibid., p. 23.

23. Ibid., p. 18.

24. Public Law 95-217, Sec. 202 (1978).

25. Maryland Department of Health and Mental Hygiene, Office of Environmental Programs, *Sewage Construction Grants and Planning Programs, A Status Report on the Application of Innovative and Alternative Technologies in Maryland,* 1980, p. 1.

26. *Annotated Code of Maryland,* 43:387B; 43:428-444.

27. U.S. Advisory Commission on Intergovernmental Relations, *Tax Capacity in the Fifty States: Supplement, 1980 Estimates* (Washington, D.C.: June 1982), pp. 12, 19.

28. Clifford Johnson, Maryland Office of Environmental Protection, January 8, 1979, as reported in Gardner, "Intergovernmental Fiscal Relations," p. 88.

29. Gardner, "Intergovernmental Fiscal Relations," p. 118.

10 Funding Clean Water in Michigan

Doris Van Dam

Wastewater in Michigan

Michigan prides itself as being the "Water Wonderland." Virtually all of Michigan is in the Great Lakes basin. Most of our inland waters as well as every bit of sewage, industrial waste, and everything else we put into our water anywhere in Michigan eventually finds its way into the Great Lakes. Not only does everything find its way there, but for all practical purposes it stays there too since there is very little flow out of the Great Lakes. Even with water pouring over Niagara Falls, it takes fifteen to twenty years for a complete change of water in Lake Erie, and nearly a century for the same cycle in Lake Michigan.

For the first half of the present century, virtually all the construction of municipal waste treatment plants in Michigan—and elsewhere—was financed by local units of government. The interruption of construction necessitated by World War II created a substantial backlog of construction needs and resulted in a general worsening of pollution. For a variety of reasons, many communities were experiencing difficulties in financing plant construction, a situation that prompted Congress to authorize the incentive grant program of 1956.

The 1956 federal Water Pollution Control Act established a general citizen obligation to abate municipal water pollution. It provided for a program of grants to municipalities equal to 30 percent of the cost of treatment plants and interceptors. In Michigan, 138 sewage treatment projects were funded under this original program, totaling $25.7 million worth of construction. In 1966 the federal program was substantially revised through the passage of the Clean Water Restoration Act.

In this act Congress recognized the need for an increased financial commitment on the part of the federal government, and it authorized a national expenditure of $3.55 billion for grants over a four-year period (FY 1967–1971). The provisions of the grant program were revised to provide for grants of 50 percent of the costs of projects (plants and interceptors) if the

The author acknowledges the assitance in preparing this paper by Beverly Lehr, Michigan Department of Natural Resource; William R. Kelley, Michigan Department of Health; James Townley, of Fishbeck, Thompson, Carr & Huber, Inc.; and Philip Moore, Clifton Albers, and Mary Goodwin, all of the city of Grand Haven, Michigan.

139

state would assume 25 percent of such costs. Without state financial participation, the federal grant was restricted to 30 percent of construction costs. The 1966 act also provided that either the state or local governments could prefinance all or a portion of the federal share and would subsequently be reimbursed. The authorization for that federal grant program expired at the end of FY 1971.

Michigan's first state program was enacted by the 1966 session of the legislature. A Water Pollution Control Fund was created to supplement the existing federal program. The state program, like the federal program, provided a 30 percent grant. The state program was available to those municipalities on the annual priority list for federal aid and for which there were not sufficient federal funds. Once the annual federal allotment had been allocated, an additional number of projects were certified for state assistance. The number of communities receiving state aid depended on the annual amount appropriated by the legislature ($2 million the first year).

With the exception of the initial 1966 effort, Michigan's major program to abate pollution from municipal sources by way of financial assistance began in 1967. At the time the following steps were taken:

An inventory was completed of facility needs to control pollution from municipal sources throughout Michigan—this was predicated on a level of secondary treatment for a statewide 1980 population. Estimated needs were: 210 new treatment facilities; improvements to 126 existing treatment facilities; and the needs for interceptors, collecting sewers, and combined sewer overflow control systems were estimated.

Cost estimates were prepared to accomplish the needs.

An assessment was made of federal programs available to help meet these needs. Treatment needs (including interceptors) were projected to have a total cost through 1980 of approximately $568 million (including an inflation factor).

It should be pointed out that at that time secondary treatment was considered to be the best level of treatment available as the technology for phosphorus removal was not yet proven. Costs for collecting sewers and overflow control were estimated at $561 million.

Of the $3.55 billion authorized in 1966 for the federal program, Michigan's allotment was $146 million. It appeared that the authorized federal program would handle about 25 percent of Michigan's needs.

This information was summarized and presented to the governor and key legislators. Based on the conviction that Michigan must move rapidly ahead to control pollution, a determination was made to establish a major state financial aid program. Toward this end, the following steps were realized in 1968:

> The state construction grant program (Act 329 P.A. [Michigan, Public Acts] of 1966) was amended to provide for a 25 percent grant of that portion of project cost eligible for a federal grant. This was Act 75 P.A. 1968.
>
> The legislature adopted measures to place the question of issuance of general obligation bonds for $335 million on the November 1968 general election ballot.
>
> Overwhelming approval (70+ percent) of citizens of Michigan to issue Clean Water Bonds.

Two pieces of enabling leglislation were then introduced by a Joint House-Senate Interim Committee that had been meeting since early fall 1968 to prepare bills that would assure the best expenditure of the bond money. Act 21 P.A. 1969 and its companion Act 159 P.A. 1968 established the framework for distribution of the $335 million clean water bond money; $285 million was authorized for treatment facilities, and $50 million to aid in collecting sewer construction. The collecting sewer program was intended for those communities that had a water pollution problem which could only be solved by building a collecting sewer system and that would face serious difficulty in financing such a system.

These acts provided that the Michigan Water Resources Commission would have the main responsibility in the administration of the program. The commission had been previously designated by the legislature to administer the federal and state water pollution control grants. The basic element of the program was an annual list of eligible projects arranged in priority order by the Water Resources Commission and subsequently approved by the legislature. The state program was further amended to eliminate any state advance of the federal share after July 1, 1971 because congressional authorization of the reimbursement of such advances expired on that date. The Michigan Sewer Construction Fund's unobligated balance was transferred to the Michigan Water Pollution Control Fund in accordance with Act 215 P.A. of 1972, which was an amendment to Act 329. Implementation of P.L. 92–500 (1972 Clean Water Act), which included provisions for funding eligible sewer construction under the federal grant program, eliminated the need for a state program to fund collecting sewers.

The bond fund had contributed to significant accomplishments in water quality improvements. The facilities funded, along with the clean-up efforts of industries, have produced significantly cleaner Michigan waters over the last few years.

Water recreation opportunities, as well as sport fishing, are expanding in the populated areas of the state. Some of the achievements in municipal waste treatment that have contributed to these benefits are reflected in the following general figures: more that 650 local units of government have received grants from the bond fund; more than 1000 communities have participated in funded projects; and at least half the population of the state is estimated to be served by facilities funded in part by the Clean Water Bond Fund. Of the estimated 3.2 million people in Michigan currently sewered and living outside of the Detroit metropolitan waste treatment system, 80 percent are now served by plants with secondary treatment capabilities or better. An analysis of the discharges of forty-three major treatment plants between 1965 and 1976 has shown that although the volume of wastes treated increased by 34 percent, the BOD load discharge to Michigan waters by the plants decreased by 20 percent.

Despite the many projects funded, a few events have caused the bond fund to fall short of funding the facilities now estimated to be needed. The passage of the federal Water Pollution Control Act Amendments in 1972 changed the federal grant program significantly, making collecting sewers, design, and planning eligible for grants. The state grant program has paid a portion of eligible costs as defined by federal requirements and thus has paid toward costs not originally anticipated. While some sewers were eligible under the original state program, they became more widely eligible in the adjustment to accommodate the federal program. Waste treatment requirements became more stringent with the result that more costly treatment facilities are now required for some communities than had been anticipated. In addition, the complicated new requirements of the grant program under the 1972 amendments caused a two-year hiatus in construction of facilities and came to be the limiting factor in how rapidly treatment facilities are built. The complicated requirements have also made it much more expensive for communities to do the required planning.

Federal grant participation was increased to 75 percent of eligible costs, prompting a decrease of state participation from 25 percent to 5 percent. These savings, however, have been more than wiped out by the changed pollution control requirements, the delay in construction, and the unforeseen high inflation rate. Indeed, the inflation rate for some years was more than double the anticipated rate.

By 1982 Michigan had awarded grants that surpassed the cash available from the $302 million in bonds that have been sold since 1972. However, since 1972 the interest rates at which tax-exempt bonds are sold on the national market have increased to the point that the state is now precluded from selling the remaining bonds for the purposes authorized by the voters in 1968. Although the questions submitted to the voters at that time did not specify any interest rate, the act that was passed by the legislature and signed by the governor referred to a 6 percent maximum rate.

In view of the practical limitations of the 6 percent maximum rate, Act 47 of the Public Acts of 1982 was enacted into law on March 19, 1982, with an effective date of September 19, 1982. This act increased the 6 percent interest ceiling on the sale of clean water bonds to 18 percent. The bond consulting firm for the state would not underwrite selling of the bonds at a higher interest rate unless it was ratified by a vote of the people or confirmed by the state supreme court that it is legal. Therefore, the Michigan Supreme Court has been asked by the governor to reveiw this act and determine whether Act 47 of the Public Acts of 1982 is legal without specific voter approval of the increased interest rate and whether it may be relied upon by the state as the basis for borrowing the additional necessary funds. The inability to augment the Clean Water Bond Fund will require local governments to finance projects out of their already limited financial resources. It is possible that local governments, which would be unable to finance the additional 5 percent cost, would lose the 75 percent matching federal funding.

Currently [November 1983] the state has balanced its budget with $358 million in cuts and layoffs and higher taxes on income and cigarettes. The deficit, however, is $615 million and the state work force was reduced by 10,000 in the past eighteen months. The economy is in a near-depression and state revenues are down, mainly because of the automaking and housing construction slumps. Indeed, the fiscal climate in Michigan is so dismal that at this time plans to pursue alternative state programs for the financing of projects have not been formulated.

It should be emphasized that Michigan has still existing backlog needs, that is, those capital improvements necessary to accommodate the population in existence in 1980. (By contrast, *total needs* are those needs for capital improvement to accommodate current population and projected growth to the year 2000). In late 1981 Michigan was listed at 4.1 percent of the total national needs and 4.7 percent of the national backlog needs. At that time it was considered that Michigan's total needs were $4,958,786 and that national needs were $119,893,000. Michigan's backlog needs were $4,246,030, and the national backlog needs were $90,892,000. The Michigan Department of Natural Resources is currently working to update that data and provide more accurate figures to estimate Michigan's projected growth needs.

Wastewater Situation in Grand Haven, Michigan

The city of Grand Haven and village of Spring Lake are located along the south-central portion of Mighigan's lower peninsula's Lake Michigan shoreline. The topography is generally flat to gently rolling, with some moderately hilly areas along the dunes of Lake Mighigan. The area receives precipitation in all seasons, and is relatively well drained by the Grand River. The economy of the area is linked to industry, agriculture, and recreation. The first of these is currently declining, the second is holding its own, and the last is actively increasing. Its residents make full use of the area's recreation opportunities, especially its waters, and show every evidence of wanting to keep these waters better than average in quality. Grand Haven has a population of 11,763 (1980 census), and the area has a population of 37,583. In 1967 the Michigan Water Resources Commission ordered phosphorus and other pollutant reductions from communities and industries discharging wastes to the waters or tributaries of the Great Lakes. The commission called for an 80 percent removal of phosphorus from Grand Haven, Spring Lake, and Eagle Ottawa Co., as well as combined maximums for all three sources of 1,126 pounds per day of biochemical oxygen demand and 30 mg/l of suspended solids. At the time Grand Haven operated an outdated primary treatment plant, and Spring Lake used an overloaded Imhoff tank. The leather company effluent from the Eagle Ottawa Leather Company, a principal employer, was simply screened before discharge to the Grand River. Tri-cities area (Grand Haven, Spring Lake, and Ferrysburg) water quality problems were also exacerbated because of a high water table, sand mining, and the fact that 10 percent of the area is in impermeable surfaces.

In response to this situation, in 1970 the city of Grand Haven and the village of Spring Lake entered into an agreement establishing the Grand Haven-Spring Lake Sewer Authority to acquire, own, improve, enlarge, extend, and operate a sewage disposal system in accordance with the authorization contained in Act 233 of the Public Acts of Michigan for the year 1955, as amended. Since that time the authority has become the regional unit to provide wastewater treatment services to the entire area surrounding Grand Haven and Spring Lake.

The treatment facilities, which became operational in 1973, were constructed by the authority through the issuance of authority bonds and with the aid of federal and state grants. The city of Grand Haven and the village of Spring Lake contracted with the authority to make annual payments for principal and interest to the authority sufficient to retire the 1972 authority bond issue. The bond issue was for an amount equal to $1,925,000 and will expire October 1, 1982. Spring Lake Village paid the 1972 bond debt retirement fund with user charges until 1980. Since that time it has been paid for by a special millage levy.

In late 1982 the city of Ferrysburg and Spring Lake Township—pursuant to the agreement entered into in 1972 concerning access to the plant—exercised their right to purchase the unused capacity in the wastewater treatment facility. They requested that such unused capacity be made available for their use on a first-come, first-serve basis until such time as such unused capacity is needed by the city and/or the village. In addition, the township and Ferrysburg purchased capacity interest in portions of the plant that were oversized to facilitate future expansion of the plant at a later date. Each purchased one-third of the 1.5-mgd capacity of the portions of the plant that had been oversized for $225,740. That payment was divided by the city and village on a 93 percent-7 percent basis.

If the authority felt it necessary to make capital improvements to the plant or acquire additional land later, then the cost of such improvements or acquisition would be allocated between the city, village, Ferrysburg, and the township. Each would pay, in cash, their share of the cost of such improvements or acquisition at the time of payment unless the authority financed the cost, in which case each would pay proportionately all debt service payments required to amortize the indebtedness incurred to finance the project.

Each unit pays costs for operation and maintenance based on the separately metered flow entering the treatment plant. Other rights and obligations of the parties are also spelled out in detail.

The Eagel Ottawa Leather Company is a major employer in Grand Haven and a generator of substantial amounts of wastewater. In 1970 the city and the leather company entered into an agreement to provide for cost sharing in the constructing, maintaining, and operating of the new wastewater treatment plant to be constructed by the authority.

The leather company shares in the operating and maintenance costs of the plant, based on its volume, biochemical oxygen demand, and suspended solids. Its share of the cost of the new plant was determined by mutual agreement of two independent professional engineering consultants, one designated by Eagle Ottawa and the other designated by the city.

Clean Water in Michigan

In general, Michigan's water supplies are in good shape, and the state has excellent, from a health standpoint, groundwater supplies. Historically, the public water supplies have been excellent water sources, and over the past twenty years they have seen little degradation; in fact there has been some improvement since 1972.

During the past eight to ten months, groundwater supplies in Michigan have been monitored for trace organics. Perhaps a dozen problems surfaced, none that were life threatening, but those that were found were generally controlled by taking the affected wells off the system. The Michigan state government works with communities that have problems, and Public Act 399 of 1976—referred to as the Michigan State Drinking Water Act—deals with this.

The American Water Works Association, its Michigan chapter, and the

Michigan Department of Public Health has gone on record as wanting no construction grants program. For ten years the Farmers' Home Administration (FHA) has had a program for small communities that has worked very well. Early on, the FHA attempted to impose their standards along with the grant monies but they no longer do so, and in fact, they have worked very cooperatively with the state. A review of the Grand Haven situation will illustrate how one community is grappling with the problem of expanding its pure water facilities.

Grand Haven Fresh Water Treatment

The original source of water of the city of Grand Haven was thirty-three wells located along the Grand River and Lake Michigan. In 1912, two one-half-million-gallon storage tanks and a steam-driven pumping station were constructed. The existing water filtration plant was constructed in 1927. In 1950, the treatment plant was taken out of service and a Ranney well system was installed that uses Lake Michigan as a water source. A new pipeline was installed, and the thirty-three wells were abandoned. The Ranney system is still in service today, and water from the Ranney collectors is pumped into a chlorine retention tank and directly into the distribution system.

By 1961 water demands could not be met, so a submerged intake was installed. Water from the submerged intake is prefiltered on the bottom of Lake Michigan, pumped through the water treatment plant where it is filtered and then sent through the distribution system. In 1975 the city signed a water agreement with Ottawa County Road Commission to construct a new North Ottawa County water facilities system. Engineers were hired and studies undertaken to service six nearby governmental units in addition to Grand Haven.

Studies conducted in 1978 determined that the city's water demand would exceed 11 mgd by 1981, and it was then concluded that it was necessary to expand the existing water supply capacity from 8 to 11 mgd as quickly as possible. At the time it was thought that the higher treatment rates through the plant would require replacement of the existing sand media with a mixed media to allow for high filtration rates. This same study also pointed to the fact that two hours of detention, as required for disinfection, could be provided by the existing treatment facility. A filter rate of 5 gpm per square foot of filter bed would be required to process 3,500 gpm through the existing plant. A meeting between the city's consulting engineers and the Michigan Department of Public Health (MDPH) resulted in the MDPH giving temporary permission to allow the present treatment facility to operate at the 3,500-gpm flow rate until the new facility is constructed, as long as the higher filter rates are only used for peak-demand periods.

The raw water supply was expanded approximately 3 mgd in the winter of 1979 by installing new pumps to increase the pumping capacity of the submerged intake pumping station. The treatment plant is able to hydraulically handle upwards of 3,500 gpm, or approximately 5 mgd. Since the Ranney intake can handle 6 mgd, it is possible for the city to supply 11 mgd.

This was only a temporary solution designed to satisfy short-term water demand, however, and it was still necessary to provide for the anticipated water demands for the service area through the year 2000. The objectives were to provide high quality water at the lowest feasible price. Toward this end, a liminology study was performed in 1977–1978 and it demonstrated that the location of the existing intake system is undesirable due to poor water quality and ice problems. Because of its closeness to shore, the capacity of the existing intake system has been restricted during the winter season due to the grounding of surface ice that extends down to the lake bottom. Because of its relative closeness to the Grand River outlet, the water quality in the vicinity of the existing intake system is less than desirable as a raw water source. The preferred location for a new raw water intake would be 1½ to 2 miles south of the Grand River outlet and ⅓ to ¾ miles offshore, depending on the type of intake used.

With the liminology study complete, a detailed engineering study was undertaken to determine the most cost-effective intake and treatment system. Two types of intake systems were evaluated, namely, submerged and crib. Cost estimates showed that the submerged intake was the most economical of the two. Furthermore, due to prefiltering, the quality of water obtained from a submerged intake is better than that from a crib intake.

Given the high quality of water that would be obtained in the submerged intake, a direct filtration treatment plant was proposed rather than a conventional treatment system. With a direct filtration plant, the clarifiers that normally precede the filters are eliminated, and all solids are removed on the filters. This drastically reduces construction costs as well as operating and maintenance costs.

At the time the direct filtration system was proposed, no direct filtration plants had been approved by MDPH for construction in the state. However, after careful documentation and consideration, the MDPH agreed to the direct filtration concept for this plant. The total capital costs for direct filtration has been estimated at $11,197,000, and the corresponding operation and maintenance costs are estimated $258,000 per year. These cost estimates include the costs of the intake, raw water pumping, raw water transmission main, treatment, and finished water transmission main. They do not include any costs associated with the distribution system.

For the purpose of arriving at a projected annual debt payment, the 1979 study evaluated financing the project by 8 percent revenue bonds amortizing over some thirty years and general obligation of the same type.

The analysis showed that the overall unit cost for the city's water system could increase an estimated 120 percent under the general obligation bond method of financing, or 160 percent if the revenue bond approach was selected.

Three management alternatives that could be used in the financing and administering of this project are: (1) management and bonding by the city of Grand Haven, (2) management by the city of Grand Haven and bonding by an intermunicipal authority (that is, government units included in the Grand Haven service area), and management by the city of Grand Haven and bonding by Ottawa County. When using revenue bonds, any one of the three management alternatives may be used.

If the general obligation bond approach is used, it is highly unlikely that the city of Grand Haven would have the bonding capacity required to pledge its full faith and credit. On the other hand, it appears likely that the added tax base from an intermunicipal authority would have the bonding capacity needed to pledge its full faith and credit behind a general obligation bond issue. The third alternative of selling bonds through the county has been used a number of times in the past for selling general obligation bonds, and it is likely that this method may also be implemented for the project reviewed in this study.

After all the participants in the project reviewed the preceding information, it was decided that design of the proposed project should be performed concurrent with formulation of a financial plan. After some delays in site acqisition, the design was authorized in April 1980. The design was completed in January 1981, and MDPH subsequently issued construction permits.

At an early meeting it was decided that the city and the country each would finance their own share of the project separately. The city decided that it would fund the project with revenue bonds since that would not effect the city's general obligation bonding capacity for future projects and it would be more feasible and easier to implement under current Michigan laws.

By the time construction permits were issued in early 1981, nearly two years had passed since the initial engineering study was performed. During this time several factors began to affect the funding package. First, the prevailing interest rates for all types of bonds had risen dramatically, and it was no longer feasible to sell bonds for a thirty-year period as had originally been assumed. A much shorter bond period appeared necessary. In the two-year period since the report was completed, inflation—which was running at a very high level during this period—had increased the construction cost of the project significantly. In addition, 1980 census data indicated that the population in the service area was not increasing as rapidly as had originally been projected. Consequently, demand for water would not be as high as planned. This change had relatively little effect on the city of Grand Haven since it is essentially at saturation population. However, the outlying areas, particularly the townships, had counted on a rapid increase in population to create a greater number of users over which to spread the cost of financing their share of the plant.

With these factors in mind, the Ottawa County users expressed concern that they might not be able to finance their portion of the project. As a result, the project was placed on hold in hopes that the bonding market would improve.

By mid-1982 it was apparent that the bonding market would not improve substantially in the near future, yet the need for improved and expanded water facilities still remained. However, the estimated revenue necessary for debt retirement for this project had approximately doubled due to the construction cost estimate's increasing from $11,197,000 to $14,180,000; the interest rate's increasing from 8 percent to 13 percent; and the maximum bond period's being reduced from thirty to twenty years. With these factors in mind, new methods of approaching the problem were considered.

The first step was to revise the population flow estimates and to consider a design life of the proposed facilities somewhat shorter than the twenty-year life originally proposed. Based on these factors, it was concluded that a 12-mgd plant rather than an 18-mgd plant should be considered. However, even with the reduction in cost resulting from the smaller plant, the costs were still felt to be too high to be acceptable to the communities. Therefore, the current proposal is to consider a staged construction whereby a new water treatment plant would be constructed and financed in the immediate future. Then at some later date, as the existing lake intake system and raw-water-pumping facilities become inadequate, they would be replaced with new facilities. This type of an approach will meet the objectives of supplying high quality water in the necessary quantities for the immediate future and will result in water rates that are acceptable even in the current financial climate.

In conclusion, it is apparent from the recent experiences with the Grand Haven water system that planning, design, and financing water improvements must be approached differently in 1982 than they were in 1979. Growth rates are not as high as they once were. The public is no longer as willing as it once was to finance projects to serve for many years into the future, but appear more interested in meeting only their immediate needs. The public is also very concerned about any increases in utility rates and is demanding that they be kept to an absolute minimum.

Finally, it is also apparent that the staged-type construction of system improvements may once again be a reasonable alternative. For many years it was felt, correctly so, that the most economical way to construct improvements was to build the entire system at one time and to finance the system at interest rates that typically would run 5–8 percent while the inflation rate for construction of the facilities was several percent higher. However, that situation no longer exists. Inflation seems to be under 10 percent for these types of systems. Interest rates are in the neighborhood of 11–13 percent, and even though it is often slightly more expensive to construct the system in two or three steps than in one major step, doing so may now be the only way that many communities can afford to improve their systems.

11 Funding Clean Water in New York State

Richard Torkelson

Water Supply and Quality

New York State is one of the most water-rich states in the nation. Water is one of its most precious natural resources. As with many of life's precious things, it is often taken for granted.

In 1965 and 1972 New York State began a massive program to clean up its 3.5 million acres of lakes and more than 70,000 miles of rivers and streams from the visible pollutants that were threatening human health. More than $7 billion of federal, state, and local money has been spent on this effort, spawning a massive construction program to build sewer plants to remove raw sewerage, phosphates, nitrates, and other pollutants dumped into streams and other bodies of water. Approximately 1,100 municipal projects were constructed in New York to control conventional pollutants.[1]

In spite of the success that this program has had in substantively cleaning up the waters, New York is now confronted with a related but different problem. Many of the sewer plants constructed in the state are old or were designed during a time when energy was cheap. The cost of retrofitting and rehabilitating these plants will soon have to be addressed—and borne—by municipal governments. Any hope for federal assistance in this area has been dashed in light of the Reagan administration's New Federalism actions to reduce the national sewer construction fund down to $2.4 billion. Moreover, the administration has also changed their cost sharing in 1985 from 75 percent to 55 percent and will no longer cost share preconstruction costs. The net result is a reduction of federal dollars and commitments, thereby placing a larger burden on state and local governments. The final outcome of this initiative will be to end the Sewer Construction Grants Program as we have known it over the last fifteen years.

Although New York State citizens responded with approval of two full faith and credit bonds amounting to $1.65 billion during the 1960s and 1970s to allow the state to provide at least half of the nonfederal share of sewer construction costs, with the decreased federal share, these monies will be insufficient to complete the job. Staff of the New York Environmental Conservation Department now estimates that its bond funds will last only until 1990. New York State has identified additional construction needs of almost $18.5 billion in the EPA *1980 Needs Survey for Publicly Owned Wastewater Treatment Facilities.* Completion of all these projects under the current federal authorizations, assuming a zero inflation rate, would take more than fifty years. Indeed, if one only looks at those categories that are financially eligible under the Construction Grants Program as recently amended by Congress (secondary treatment, advanced secondary treatment, advanced treatment, infiltration/inflow, and new interceptor sewers), New York State needs more than $7 billion. This does not include the state's identification of more than $3.3 billion in replacement or major rehabilitation needs for existing sewer facilities.

While much attention has been paid to cleaning up visible pollutants from our water through the sewer construction program, the state has had to turn its attention also to problems arising from invisible toxic contaminants. Organic chemicals, both natural and man-made, found in the state's drinking water sources have been shown to have adverse impacts on the environment and public health. The sad situation of Love Canal has demonstrated that these chemicals, when improperly stored or discarded, will work their way into the environment and eventually into our sources of drinking water.

New York's overall water pollution picture, it must be pointed out, is also affected by the types of water sources used by its citizens. Roughly one-third of the state's population (6.3 million people) uses groundwater for their water supply, while the other two-thirds uses surface waters. When New York City's population is removed from the surface water calculation, the number of people dependent on groundwater jumps to over 60 percent of the population. Surface sources of drinking water are generally not affected over the long periods from the contamination incidents. Water flow generally cleanses surface water in a short period of time. Groundwater, on the other hand, moves very slowly. Consequently, contamination will take years or even centuries before it will move out or pass from ground sources of water. The only timely method for cleansing contaminated groundwater is through very expensive treatment.

Continuous sampling undertaken by the New York Department of Health has already identified serious contamination in numerous groundwater supplies throughout the state. On Long Island, for example, 2.5 million people are totally dependent on an underground water supply that has increasingly been found to contain a growing list of toxic chemicals. Although the scientific record is yet to be finalized on the long-term effects of these chemicals on humans, the state has already forced the closing of many water wells. Consequently, we are confronted with a new environmental health problem: the need to mandate water treatment of contaminated groundwater supplies to protect public health. The cost and determination of who pays for this treatment is as yet an unanswered question. Currently, no federal or state grant program exists to meet this need.

Water Management and Infrastructure

New York State uses approximately 3 billion gallons of water a day, or roughly 1 trillion gallons a year. There are over 1,800 water systems (source, transmission, treatment, distribution) throughout the state. Of these 80 percent are publicly owned, 20 percent are private. Built in the earlier parts of this century, many of these systems are rapidly deteriorating. Due to a lack of attention and money, pipes are breaking down, water mains are too small, water tanks are peeling and rusting, pipes are freezing, and impoundments are weakening. It is estimated that large-scale rehabilitation is needed by almost every municipality.

Indeed, it is not uncommon to pick up a newspaper from anywhere in the state and see stories of broken mains, flooding, or shutdowns of water supplies because obsolete equipment has broken down. Yet even more frightening are the volumes of water lost or unaccounted for from leakage. Unaccounted water losses range between 35 and 50 percent for most communities. With the state's recent drought experience (1980–1981), officials are necessarily concerned that conservation is an equally critical need.

Although there are no federal or state grant programs to assist in the reconstruction or rehabilitation of these antiquated water systems, the federal Corps of Engineers has assisted the state in conducting selected investigations to document water systems' deterioration and estimate the cost of repairs. Although financial support has diminished, this effort has helped the state bolster its arguments at the federal level for portions of those funds, which traditionally have been given to large, new water supply projects in other parts of the country, financed by low-interest loans made available through the 1958 Water Supply Act. The older, longer-established systems, which now need massive rehabilitation, have not received any significant attention at the federal level and are not eligible for financing through the 1958 act.

The Corps of Engineers' studies, funded from Section 214 of the 1965 River and Harbor Act, have given the state a handle on estimating its needs in this area. In its study at Manhattan—which represents 10 percent of New York City's distribution system—the corps estimated that a $90 million, ten-year construction effort was immediately necessary to rehabilitate current water distribution lines. This study also identified high-priority replacements, with the corps' recommending a more thorough network analysis to determine the full extent of the total rehabilitation needs; this early assessment of Manhattan's high-priority needs, if extended to the other boroughs of New York City, would place the cost for a ten-year pipe replacement program at $.9 billion. However, even this figure is inadequate because it does not take into account the fact that water is transmitted to New York City through two very large water tunnels, put in service in 1917 and 1936, that have never been shut down for inspection and servicing. To avoid the catastrophe of one of these tunnels's not working, the city began construction of a third tunnel system that, when finished, will allow inspection and repair of the older tunnels. Cost estimates for the new tunnel range from $2 to $4 billion, depending on how many and when its four stages are built. No one needs a lecture on the financial crisis that has affected New York City. It is sufficient to know that by being shut out of the bond market, water infrastructure needs (new tunnel and broken pipes) now compete for very hard dollars out of each year's budget—hardly the most expedient manner to address the city's water problems. Other costs to the city over the next decade will arise from the need to develop new source supplies and water treatment. In a city that needs 1.5 billion gallons of water a day, these are not casual issues.

The Corps has also studied other sections of upstate New York and documented similar needs for many older cities and rural communities. For example, it determined that Binghamton needs $8.5 million; Rochester, $200 million; Buffalo, $400 million; Watervliet, $1.2 million; and Corning and Hornell, $2.6 million.

Other studies present similarly grim pictures. In a 1981 lecture given to an American Water Works Association Workshop, J. Finck and H. Pike stated:

> Unaccounted for water generally ranges from 35–55 percent. In rural areas, this figure is higher.

> Many distribution systems are inadequate because of obsolescent materials, inadequate design, external corrosion, tuberculation, and inoperative valves.

> Leakage is causing higher treatment and pumping costs. Moreover as a result of poor system reliability and inadequate pressure, the public is paying more for fire insurance.

> Industrial development opportunities will be lost because of inadequate or deteriorating water systems. A survey by the U.S. Commerce Department of firms likely to expand or relocate showed that fire protection and water supply were critical top corporate considerations for locating their industrial plants. Other factors such as taxes and transportation access ranked below water.

> Additional supply and storage capacity is needed, particularly in rural and southeastern New York.

> Virtually all water systems lack adequate funds to begin the necessary improvements.

> Additional treatment capacity most likely will be needed to meet new safe drinking water standards for toxics. This will put an additional financial burden on all communities throughout the state.[2]

Funding: History, Problems, and Alternatives

In New York State, where most of the responsibility for water and sewer treatment belongs to municipalities, some very significant problems are building up. It is obvious that the primary problem is the gigantic cost of construction both now and in the future. A second problem is the difficulty of trying to borrow money in a marketplace where high interest rates make the cost of long-term indebtedness prohibitive. Third is the negative financial condition of many of the older, larger cities that would keep them out of the marketplace even if interest rates were low. A fourth problem confronting many cities—even those in reasonable financial shape—is the staggering need for investment elsewhere in their infrastructure (for example, roads, bridges, transit systems, and ports). Fifth, the state constitution blocks cities from revenue bonding their water and sewer infrastructure needs without going through the cumbersome process of a referendum. Finally, other constitutional provisions limit municipalities' ability to incur capital debt.

Clearly, financial help is needed. The state has and will continue to make concerted efforts to try to persuade national policymakers that it is time to develop a program of rehabilitation to take care of the water infrastructure developed over the last hundred years. It is our belief that the problems evident in the Northeast will very soon be experienced by other parts of the country. The water system is different from other infrastructure systems only in that it is not very visible to the public. Yet whether it's visible or not—be it highways, bridges, rail systems, or water systems— when infrastructure is allowed to deteriorate, it quickly becomes unusable. If we fail to invest now in these critical systems, we will have to act in crisis—and we cannot afford to wait that long in the water arena. Indeed, to give but one example, the failure of one of New York City's currently existing tunnels would, according to one study, require a massive evacuation of the city.[3]

In response to these studies, the state has developed a legislative proposal to create a state Water Finance Authority, a quasi-public agency empowered to issue up to $4 billion of debt for water and sewer improvements. The legislation provides a revenue-bond-financing capacity for all municipalities. The authority would require that all user fees be fixed by an independent, statutorily created local water board and that they be sufficient to pay debt service, operating costs, and other agreed-upon costs of the facilities. The operations and maintenance of the facilities will remain a responsibility of the municipality. The water authority proposal will require, for the first time, that adequate funds be generated to properly pay for these services (rather than having them subject to competition in the general budget every year). It is the intent of this legislation to get those participating municipalities on a user pay basis to take care of their own facilities.

Without this legislation, cities would continue to be constitutionally prevented from dedicating their water and sewer revenues to assure revenue bonds for their infrastructure and capital needs.

In New York, water is extraordinarily cheap (see table 11-1). In New York City, for example, the rate per thousand gallons is $.70; in Syracuse, $.62; in Albany, $1.00; in Binghamton, $1.13; and in Mount Vernon, New York, $.47. If additional infrastructure is necessary, it would appear that the users in the state should be able to pay for it from increased water fees. Although municipalities throughout the state (exclusive of cities) could increase their user fees to produce the necessary revenue to defray additional capital water expenditures, they must balance this new debt within the overall debt restrictions of the state constitution.

The Water Finance Authority proposal also includes very important program elements. It creates a state Water Resources Council that will be required to adopt a statewide and regional water management strategy. These water strategies are to be developed so as to reflect a reasonable breakdown of the state's water source and supply service areas. Broadly defined, they will lay out the allowable and necessary developments in a priority fashion for each of the watersheds to ensure that residential and business users will be guaranteed a sufficient water supply Moreover, these strategies would become critical benchmarks to ensure that the state is not supporting, through the finance authority, any developments in the water area that may ultimately harm its communities. Upon adoption by the council, these management strategies will then be the framework within which permitting by the state's regulatory agencies and financing by the Water Finance Authority must take place.

Although the New York State Department of Health estimated the system's needs of water purveyors at $6.8 billion in 1980, (see table 11-2), the Water Finance Authority proposal is but a beginning in the state's effort to provide a financial mechanism to meet its total infrastructure needs. There appears to be no doubt that in the future, water and sewer capital needs will have to be paid for by those who use the systems. If federal help is to be realized at all, it will probably be in a form that will limit the cost of water and sewer services to consumers.

As of November 1982, the Water Finance Authority is still not in law (it has passed only one house of our state legislature). Upon enactment it will still have to develop methods and procedures for financing requests from the many communities throughout the state. This process will take some time before the authority can be reasonably expected to approach the marketplace for its first financing. A critical hurdle in the entire scheme will be the willingness of the public to pay more for a reliable resource they have come to enjoy and expect to be cheaply provided.

Table 11-1
New York State Water Systems, 22 Largest, 65 Percent of Population

System	Type	Population Served	Total Gallons Used per Year (Billions)	Current Basic Rate per 1,000 Gallons (Dollars)	Gross Annual Revenues (Millions of Dollars)
New York City	Municipal	7,200,000	511.00	0.70	186.00
Buffalo	Municipal	400,000	38.00	0.70	15.90
Yonkers	Municipal	220,000	11.00	1.04	5.70
Syracuse	Municipal	200,000	8.70	0.62	4.00
Utica	Municipal	131,000	7.30	0.60	5.00[a]
Albany	Municipal	115,000	8.80	1.00	4.50
Troy	Municipal	110,000	6.30	0.90	3.10
Schenectady	Municipal	100,000	6.30	0.59	2.77[b]
Niagara	Municipal	100,000	7.20	0.75	3.97[a]
Mount Vernon	Municipal	75,000	4.20	0.53	1.59[a]
Elmira Water Board	Municipal	70,000	3.45	1.42	2.35
Binghamton	Municipal	65,000	4.50	1.13	1.38
Tonowanda Water District #1 & #6	Town	110,000	4.80	0.80	2.86[b]
Latham Water District	Town	82,000	3.60	0.73	1.95[b]
Erie County Water Auth. (two districts)	Authority	500,000	16.50	1.31	19.30
Monroe Co. Water Auth. (two districts)	Authority	550,000	20.00	1.11	13.34[a]
Onondaga Water Auth.	Authority	116,000	6.00	1.21	7.67[a]
Jamaica Water Supply	Private	650,000	28.80	1.41	21.15[a]

Table 11-1 (continued)

System	Type	Population Served	Total Gallons Used per Year (Billions)	Current Basic Rate per 1,000 Gallons (Dollars)	Gross Annual Revenues (Millions of Dollars)
Long Island Water Corp.	Private	260,000	8.80	1.40	12.50
Spring Valley Water Co.	Private	241,000	10.10	2.38[b]	23.30[a]
New York Water Service	Private	170,000	5.00	1.83	6.54[a]
New Rochelle Water Co.	Private	150,000	8.00	1.75	7.98[a]
Totals		11,615,000	728.35	(Avg.) 1.09	352.85

[a]Projected annual revenues based on recent increase in rates.
[b]On December 30, 1981, Spring Valley rate increased to $2.971 per 1,000 gallons.

Table 11–2
Water System Funding Needs

Category	Billions of Dollars
New York City water supply	4.2
Other urban water supply	1.443
Rural water supply	1.156
Total	6.799

Source: Adapted from New York State Department of Health *1980 Needs Survey.* Survey responses do not necessarily conincide with needs identified by Corps of Engineers infrastructure studies.

Looking into the future, the survival of the state of New York will depend on the reliability of its water and sewer infrastructure. Major new industries, such as food and beverage processors, chemical, steel, lumber, wood, and paper manufacturers require abundant water resources. Consequently, the economic development aspects of this resource cannot be ignored. Although the work necessary to complete all our construction needs will not be done overnight and may even extend beyond the year 2000, we can expect that by moving now we can not only retain our industrial base but also reattract some of those industries that moved to other parts of the country because of cheaper labor and energy costs. If the Water Finance Authority successfully gets off the ground, it will become the major link in ensuring a viable economy for New York in the future.

Notes

1. New York State Department of Environmental Conservation, *A Challenge for the 80s* (Albany, 1982), p. 17.
2. J. Finck and H. Pike, "Infrastructure Rehabilitation—Where Do We Go from Here?" (Paper presented at the American Water Works Association Workshop in Rochester, N.Y., 1981), pp. 9–10.
3. General Contractors Association of New York, Inc., "Imperiled Lifeline: Prospects and Problems for New York City's Water Tunnel No. 3," *Public Policy Report,* No. 1 (1983), pp. 8–12.

12 Funding Clean Water in South Carolina

Clair P. Guess, Jr.

In general terms, public attitudes seem to be that wastewater treatment is being well cared for once a treatment plant comes on line. This assumption is misleading, erroneous, and at best only partially true. The public spoken of here has limited knowledge of initial cost of construction, maintenance and operating costs over time, plant efficiencies or deficiencies, and cost-benefit analysis. Importantly linked is the question, Do nationwide standards, rules, and regulations established by the Environmental Protection Agency (EPA) apply reasonably in all sectors of the nation? Who, then, will bear the cost? The public often pays much more than it should for potable water through increased taxes and user charges and through the purchase of consumer goods, influenced upwards by the cost of water treatment.

Water Supply and Quality: A General Accounting Office View

In response to a congressional inquiry made by Congressman Norman Y. Mineta, chairman, House Committee on Public Works and Transportation: Oversight and Review Subcommittee, the General Accounting Office (GAO) had this to say through a release of December 15, 1980:

> GAO found that over $25 billion in Federal funds and several billion dollars in State and local monies have been spent to construct new wastewater treatment plants or to significantly modify existing plants. The EPA estimates that through the year 2000 an additional $35.6 billion in Federal funds alone will be needed to construct additional treatment plants. GAO found that many of the plants, in operation for several years, have seldom or never met the performance standards they were designed to achieve. Failure of treatment plants to meet performance expectations may not only have an adverse impact on the Nation's ability to meet its clean water goals, but may also represent the potential waste of tens of billions of dollars in Federal, State and local monies.[1]

This GAO review also points to properly designed plants with high technical sophistication that were very expensive to operate, so much so that the local authorities either bypassed the operation or reduced operating and maintenance costs to a level that made the operation totally ineffective. Of course, a limited number are reported as being maintained and operated at acceptable levels.

Chlorination: A Hazardous Treatment

Another serious challenge to wastewater treatment is the use and excessive use of chlorine in the treatment process. Dr. James B. Coulter, secretary of the Natural Resources Department of Maryland, has pointed out:[2]

> Chlorination of ordinary sewage treatment plant effluent provides no significant public health protection and to the contrary it can result in public health hazards that go undetected.
>
> At very low concentrations, chlorine is acutely toxic to fish and other mature forms of aquatic life.
>
> At much lower concentrations, chlorine decimates first emergent forms of life (such as fish larvae).
>
> At much lower concentrations, barely detectable with the most advanced analytical techniques, chlorine and its organic by-products repel migrating anadromous fish so that they are denied access to spawning grounds essential to their propagation.

Dr. Coulter points to alternatives to chlorination, each of which have their problems, while maintaining that present levels of disinfection may not, of necessity, be a realistic goal.

New or Modified Concepts of Treatment

Water is a key element essential to survival. To harness its multiple use and reuse is essential. We cannot afford to continue to employ methodologies that are less than fail-safe. We should employ a worldwide sharing of technology in the field of wastewater treatment, and it is essential that we do this nationwide.

One outstanding product of research investments made by the federal government through the Office of Water Research and Technology, Department of the Interior, is the reverse osmosis system for water treatment. The equipment for this was an answer to an intensive search for a process that would remove salt from saline waters. This process has a wide range of adaptation, depending on varying designs and composition of the filter media. The cost of this method varies widely, depending on several factors, of which intake qualities are controlling. Prices for finished potable water are reported to be in the range of $1.30 per thousand gallons in terms of economic values in the decade of the 1960s. During this same period, industry, as a rule of thumb, felt that 30¢ per thousand was a breaking point in its favor for high-volume usage.

Another process for wastewater treatment, perhaps best suited to effluents of 2 million gallons per day or less, is spray irrigation. In this process there are more variables, including slope, ground cover, soils permeability, rainfall, and evaporative influences. At best, however, it appears that an average of 6/100-acre disposal area per capita, or for a community of 5,000, approximately 300 acres, seems wasteful of land use although it may be less expensive than tertiary treatment at the outfall.

Toxic and nonbiodegradable water pose quite different purification problems. An all-out quest for elimination of this type of waste must become and remain a top national priority. There is no better place to clean up this type of waste than at its source, in-plant or on-plant sites. It is ridiculous to allow traces or large volumes of toxic wastes from known sources to combine with other less dangerous organic wastes. To do so means excessive treatment of large mixed volumes, higher costs, and often poor treatment because of the wide range of toxics and inadequate technology for treatment at the final effluent.

Wastewater Synopsis, South Carolina

The South Carolina Department of Health and Environmental Control is the agency responsible for water quality, solid waste disposal, and other health and environmental affairs. This department is singularly responsible to the EPA in meeting national criteria, standards, guidelines, and directives related to environmental quality.

A recent South Carolina Department of Health and Environmental Control report of eighty-eight waters examined is an important commentary as background to overall progress in improved water quality for the state. It states, "Overall trends for water years 1979, 1980 and 1981, using the Water Quality Index, showed three waters improving in quality; 58 showed no trends in quality; 18 showed a lowering in quality, and for nine insufficient data to detect a trends."[3]

This report points to a nationwide expenditure of $120 billion to meet the requirement of the Clean Water Act by the year 2000 for publicly owned water treatment works. Of this national total, South Carolina estimates are $871 million, or a per-capita cost of $232 based on a projected population of 3,122,000.

South Carolina Water Use-Demands

As further background, a summary of use-demand is helpful, to distinguish areas of major concern and areas of lesser concern in wastewater treatment, pricing, and market values for water supply and wastewater treatment.

An aggregate use-demand of 9 trillion 40 million gallons (9,040,000,000) of water per day is the estimated projection of the South Carolina statewide needs for the year 2020. In addition, not included in the aggregate is a daily demand of 64,647 million gpd for hydropower, which flows through undiminished. This use-demand analysis is from a section of the prepublished State Water Plan nearing completion by the South Carolina Water Resources Commission.

Interestingly, this draft report of the commission delineates between categorical use and source. For all uses, 8 percent will come from groundwater and 92 percent from surface waters. A review of the draft report reflects the opportunity of use and reuse of these waters as they flow to the sea. The exceptions are losses due to high evaporative requirements of thermoelectric plants' use of cooling towers (or equivalent) imposed by the EPA in order to eliminate or reduce thermal loading of streams, lakes, and reservoirs. Agricultural irrigation is another use-demand that is considered 100 percent consumptive.

For South Carolina's daily-use factors for 2020, 73 percent (6,593,000,000 gpd) is for thermoelectric; 11.9 percent (1,077,000,000 gpd) for agricultural irrigation; 7 percent (636,000,000 gpd) industrial self-supplied; and 6.87 percent (620,648,000 gpd) for public supply including mixed domestic-industrial operations. The lessor uses include 1.04 percent (94,556,000 gpd) for rural domestic uses, all from groundwater. For agricultural livestock, only 19/100 percent (17,000,000 gpd) is provided mechanically from ground and surface waters. A much larger quantity is provided by streams, ponds, and dug watering holes not included.

Significantly, the major burdens of wastewater treatment within the preceding categories are industrial self-supplied, 7 percent (636,000,000 gpd) and public supplied 6.87 percent (620,648,000 gpd). These two categories, for the most part, must treat intake supplies as well as effluents (discharges) in order to meet the water quality standards of the EPA administered by the South Carolina Department of Health and Environmental Control. Thus approximately 14 percent (1,256,648,000 gpd) of the daily use-demand is intensively regulated, permitted, inspected, and under constant operational management.

A second category having to meet very significant treatment requirements is thermoelectric use. This category for South Carolina accounts for 73 percent (6,593,000,000 gpd). Treatment on the intake side is largely determined by the industry in meeting their mechanical process requirements. On the discharge side the chemical or bacterial treatment is usually insignificant with the exception of fly ash and minor periodic blow-down of condensates. A very major factor, and one that adds a large amount of cost to the electric consumer, is the cooling of effluent waters from steam to within 5 °F of ambient temperatures in the receiving waters. Usually this cooling is accomplished through forced draft towers or modified natural draft towers. Agreement on this issue remains an open battleground between the electric industries, the public demand for cheaper energy, and the EPA. Upholding the need for this treatment without modification is an element of the environmental movement.

Consumer Costs: Management and Infrastructure as Principal Needs

The cost to the consumer for wastewater treatment has, to a large degree, been based on a rule of thumb that in effect is, "take the dollar value of treated water consumption by the customer and add the equivalent amount for the cost of wastewater treatment, keeping separable accounts for each. . . ." There is merit to this formula, especially for the smaller communities of, perhaps 20,000 population or less. It is simple, easy to bill, and it has a ring of fairness to the public.

In larger communities, 50,000 to 1,500,000, complexities seem to multiply. A more sophisticated system is often used, depending on distribution investments, bonding obligations, construction costs influenced by geographic location, and other contributing factors.

The cost of potable water has, for the most part, doubled in the last decade under the principle of applying waste treatment costs to potable supplies. This seems reasonable, but to the occupants of a single-dwelling unit accustomed to $5 water cost per month, it now means $10 or more. To a large-volume user such as industry, hotels, bottling plants, and various manufacturing plants, it means, for an 8-in. meter installation having a supply of 11,200,000 gallons per month—without wastewater charges—an average of $5,934. If the effluent enters a combined collector system for treatment by a sewer authority or municipal entity, it now costs an additional average of $7,596 per month for waste treatment, or a total of $13,530 per month.

Those who pay for the services of wastewater treatment have the right to know whether or not the investment is indeed accomplishing the intended goal. They also have the right to know that local management is adequate in order to obtain the best return on investments. They have the right to know least-cost technology is being applied.

Local citizens must be on the alert to make sure that costly blunders are reduced, that research and technology are advanced, and that unreasonable treatment is not used where a lesser cost could serve the same purpose. Truly all the decisions cannot be left with a single agency (EPA) without question. Generally speaking, the environmentalists have been heard. Money for wastewater treatment is being spent by the billions, not just federal dollars but state and local as well. Combined, the sums are astronomical.

Comparison of the Costs of Water Systems

It would be wonderful to lay the income and expense data of a number of water distribution and sewage disposal systems along side each other and to be able to draw accurate conclusions about how to accomplish the best and most economical ways of operating. Unfortunately, no two water systems are alike, and the differences are so complex that this cannot now be done. While the pricing of water services is constantly in the political forefront, our ability to make useful findings about whether one utility is better operated and managed than another is typically limited to observing obviously bad practice and to making judgments about the influence of factors that may be highly technical.

Arthur Young and Company made a study for the city of Jacksonville, Florida, that well illuminates these complexities. In this study, thirty-five communities in the South were chosen for a comparison of water and wastewater pricing structures. The study concluded that the major factors impacting pricing are: geographical considerations, including customer density; demand considerations; consistency of type of customers in an area; levels of treatment; the level of subsidization, including grant funding; the age of the system; infiltration and inflow levels; and, rate-setting methodologies by which costs are distributed to classes of customers. The report concluded:

> In summary, care should be taken in drawing conclusions regarding water or wastewater operations or management in a particular community. Many factors influence water and wastewater pricing. Comparisons among communities could signal to management, however, that there should be reasons why one community's rates are higher or lower than another community's. Analysis into why there is a difference could be helpful in examining the effectiveness of a water or wastewater operation.[4]

Despite the inability of a community to accurately judge its system, except as it may rely on the opinion of experts, there are two major considerations that must be accounted for: first, maintenance and operation of wastewater treatment plants are sophisticated operations that must be funded routinely for decades beyond the initial capital investment. Second, staffing by properly trained persons with excellent technical and managerial ability is imperative if the task of treatment is to be accomplished over the long term. In this matter it should be noted that larger communities (20,000 and above) can afford more technically oriented staffs than smaller communities. Revenues from larger operations appear to be more self-sustaining, whereas the smaller ones reach points of diminishing returns.

Funding Options

In light of the foregoing, it is impossible to reasonably estimate how much will, or even should, be spent in South Carolina in the next few decades to provide clean water. It will be most difficult to prevent further deterioration of surface waters and very unlikely that "fishable/swimmable" will be attained on any large scale except where it is already present. In any case, the amount of need funds boggles the mind. Funding will have to come from federal grants and loans, state funding, and where it can be accomplished, local funding including the sale of revenue bonds and revenues from operating income. There is nothing new about these approaches. Linked with these approaches, it seems apparent that urgent attention is called for by:

Expanded research to seek the least-cost alternatives to wastewater treatment borne by the federal government

Tougher inspection, most frequently applied during the time of design and construction of treatment plants, on the part of the state and local interests

Move nearer to sources for the removal of toxic waste

Require indemnity bonds as an enforcement tool

Provide substitutes for chlorine disinfection where it has become more harmful than beneficial, creating larger problems than are solved by its use and often overuse.

Finally, HR 7781, offered by Congressman Wesley W. Watkins of Ada, Oklahoma, on July 21, 1980, would have established a National Water Utilities Bank that would have provided:

Financial assistance for the construction, maintenance, and improvement of public water systems governed by a bank board.

Low-interest-rate loans to be repaid over a reasonable period of time.

Reinvested revenues from repayment to other loan accounts

Funding by the sale of stock

The issuance of issue notes, bonds, debentures, or other obligations

Although HR 7781 failed to pass, it appears to have great merit.

Conclusion

The cost of water at all levels is predictably on the rise, even without inflationary influence. It appears to be high time to face up to our errors and
take the responsibility to see an improved system work. The alternative is
costly, cumbersome, and often malfunctioning at one level or another. In
short, dollars alone cannot do this job. We must muster additional technologies and streamline administrative procedures lest we spend ourselves
broke while pointing to others to carry the blame.

Notes

1. U.S. General Accounting Office, vol. 6, no. 1, p. 121.
2. James B. Coulter, *Barre of the Bay—Chlorinated Sewage* (Annapolis: Maryland Department of Natural Resources, 1982.
3. J.N. Knox and Larry E. Turney, *Water Quality Assessment 1979–
1981,* Office of Environmental Control Report No. 018-82.
4. Arthur Young and Company, *Water and Wastewater Rate Survey
for Major Southern Cities,* Report to the City of Jacksonville, Florida,
1981, p. 7.

13 Funding Clean Water in Utah

James O. Mason

Water Supply and Quality

Utah is the second driest state in the nation, with an annual precipitation rate of less than 15 inches per year. Utah's water quality issues and funding problems are typical of the western intermountain region, where arid conditions compel the public to give its attention to inter- and intrastate water resources and treatment for such diverse uses as culinary, recreation, wildlife, agriculture, and industry.

Water quantity and quality are the two high-priority considerations in discussions on future growth and water management, and any discussion of them must include the funding of treatment. This issue has become prominent in light of the Reagan administration's New Federalism proposal. The reduced amount of federal funds to the states to maintain environmental health programs has caused much concern about the states' abilities to maintain current program progress in water pollution control. As growth places even greater demands on states' resources, future status of water quality and quantity become even more important.

Importance of Water Treatment

Continuous intervention of the waterborne-disease cycle through disinfection of drinking water and various water pollution control mechanisms have resulted in the high standard of health enjoyed in this nation. Diseases that caused epidemics in the past are still present and more widely spread due to modern transportation systems but are held in abeyance by constant public health vigilance.

In times of tight budgets when we are reaping the public health benefits from a rigorous water quality control program in the past, it is easy to become complacent about the need for continuation of programs at the current level. However, the inescapable fact is that it requires funds to main-

The author acknowledges the assistance in the preparation of this paper of Dr. Marvin H. Maxell of the Utah Division of Environmental Health; Kenneth H. Bousfield of the Utah Bureau of Water Supplies; and Don A. Ostler of the Utah Bureau of Water Pollution Control.

169

tain programs with appropriate health standards. One does not have to look very far into the past to see the effects of even a small outbreak of waterborne typhoid fever. It has been estimated that today such an outbreak could spread to nearly 3,000 people in Utah before being controlled and that 108 people would die. Medical costs alone for one such outbreak could exceed $4.5 million. Lawsuits filed against responsible government entities may exceed several hundred million dollars. The dollars spent on prevention of environmentally caused diseases save millions of dollars.

Utah's effluent and water quality standards, established by the Water Pollution Control Committee, represent minimum levels that were determined necessary for protection of public health and preservation of designated beneficial uses of state waters. The water quality standards are formally reviewed every three years for any revisions that may be necessary. It is unlikely that substantial changes would be made in the future that would significantly reduce the magnitude of state water facility needs. Any significant reduction in standards would result in impairment of present designated beneficial uses of state waters and may increase the public health risk. In addition, drinking water regulations have been established by the Safe Drinking Water Committee.

Water Pollution Control Program

Utah's Water Pollution Control Program was formalized with the passage of the Utah Water Pollution Control Act in 1953. This act established the Utah Water Pollution Control Committee with powers to adopt regulations, set standards, review and approve plans for all wastewater facility construction, and initiate enforcement actions. Significant pollution-abatement accomplishments occurred due to this early law, and thirteen wastewater facilities were constructed during the period from 1953 to 1956.

In 1956 the federal government began providing grants for construction of wastewater treatment facilities. These grants varied in percentage of federal participation from 33 percent to 55 percent. Many communities took advantage of this opportunity to construct needed wastewater facilities to conform with the Utah Water Pollution Control Act.

The passage of the federal Water Pollution Control Act Amendments (FWPCAA), 1972, signaled a major change in the construction grants program. Federal grant participation was increased to 75 percent of eligible costs, the regulations governing administration of the program became numerous and complex, and the national commitment to funding these projects was increased. The construction grants program became the largest public works program in the country, with funding authorized at $5 billion per year. Currently the national authorization level is $2.4 billion per year.

Table 13-1

Construction Grant Funds Allocated to Utah under the Federal Clean Water Act

Fiscal Year	Federal Dollars (millions)
1972	11.0
1973	2.8
1974	4.2
1975	16.6
1976	4.2
1977	14.0
1978	20.1
1979	21.0
1980	16.8
1981	16.0
1982	11.8
Total for ten-year period	138.5

Utah's allocation of funds appropriated nationally since 1972 are shown in table 13-1. Since passage of the 1972 amendments, there have been over 18,000 grants awarded.

Since 1972 twenty-three communities with raw sewage discharges or failing septic tank systems have taken advantage of the grants program to correct these health hazards and pollution problems. Federal grants were utilized for most of the planning, design, and construction of these needed wastewater treatment facilities. During this same period, seventy-six other communities in Utah received grants for various phases of projects that ranged in scope from planning, design, and correction of extraneous flow problems in sewer systems to upgrading or replacing many existing plants. The goals of these projects have been to improve or to maintain water quality of receiving systems and to correct the development of serious health hazards.

Utah's successful record has been accomplished with $126.7 million of federal funds (1972-1981), which amounts to less than one-half of 1 percent of the amount allotted nationally for this program. These federal dollars have been matched with approximately $42 million from local sources in Utah.

At present there are eighty-nine municipal wastewater treatment facilities in Utah, and 80 percent are in compliance with effluent standards.[1] All twenty-seven facilities constructed since 1972 are in compliance with current standards.

Current Status of Surface Water Quality

As a result of cooperative interagency studies on Utah's surface waters—
under Section 208 and Section 314 (Clean Lakes) FWPCAA, 1972—the
state prioritized twenty-one stream segments where data indicated one or
more impaired beneficial uses.[2] State and local efforts are being concen-
trated to upgrade the water quality in these stream segments by proper
management of watersheds through better land use practices, regulation of
wastewater discharges, and construction of adequate treatment facilities.

Two basic concepts must be taken into consideration when dealing with
water quality problem issues that affect beneficial uses: First, water quality
in river systems tends to degrade as water flows from the headwaters to the
lower reaches of the system. The water quality of the lower reaches is a
function of the quality of the headwaters; therefore, to degrade the headwaters degrades the whole system, but degradation in the lower reaches does
not necessarily degrade the whole system. Second, water quality in a river
system is a function of the drainage through which the river flows. Just as
fingerprints identify an individual, a river's watershed basically character-
izes the quality of water that flows through that watershed. Ratios of major
cations and anions remain relatively constant, although concentrations may
vary due to a variety of weather conditions.

With these basic concepts, priorities can be established to determine the
most cost-effective measures of dealing with stream segment problems. To
illustrate the use of these concepts in Utah, the recently completed Clean
Lakes Inventory indicated fifteen impoundments with problems that impaired beneficial uses.[3] Deer Creek Reservoir on the Provo River has a
domestic water supply function as well as recreation, cold water fishery,
and agricultural function. Nutrient and biochemical oxygen demand (BOD)
loadings were imposed on this impoundment by four major discharges (two
wastewater treatment plants, a state-owned and a private fish hatchery) as
well as run-off from agricultural land and dairies that occurred upstream on
the Provo River and tributaries.[4] The two wastewater treatment plants were
closed in 1982 when a new $19 million facility operated by the Heber Valley
Special Service District began accepting sewage from the communities of
Midway and Heber. The new facility is designed to accommodate a population equivalent of 13,000, with a flow of 2.49 million gpd. Three lagoon
cells treat the wastewater, and two storage cells with a capacity of 600 acre
feet of treated wastewater provide irrigation water for alfalfa fields owned
by the district. The combined nutrient and BOD loading of the Midway and
Heber WWTP was taken out of direct discharge to the Provo River drainage just above Deer Creek Reservoir. This represents a 50 percent reduction
of the indicated loading.

The Best Management Practices proposed by the "208" studies are
being implemented for the dairies and agricultural land in the Heber Valley.

Manure bunkers are being constructed on site. Streams that run through dairy yards are being diverted or put in culverts with the necessary stockwater diverted through a series of conduit pipes to satisfy this need. Implementation of better inhouse management of fish hatchery cleaning and feeding practices has reduced nutrient and BOD loadings from these sources.

The effect of reduced nutrient and BOD loadings to Deer Creek Reservoir should become apparent in a few years. The Central Utah Water Conservancy District should not have to treat the reservoir with copper sulfate to reduce algai blooms in the future, and it should also realize the benefit by virtue of reducing treatment costs in supplying culinary water. Recreation and fishery improvements are expected.

To cite another example of wise treatment practices in Utah, in 1975 the Provo City Corporation began upgrading their treatment facility with a $20.2 million project to serve a population equivalent of 103,000 with a flow of 21.0 million gpd. The Provo Waste Water Treatment Plant discharges to a tributary of Utah Lake. The final treated water discharged by this plant is crystal clear and of sufficiently good quality to use in irrigating a nearby golf course. Faced with the prospect of expensive nutrient removal, irrigation of a golf course is definitely a cost-effective alternative.

A similar situation of wise management is being considered on the Weber River Drainage.[5] The Weber River Water Conservancy District is studying upstream uses and their implications on downstream uses. The Weber River system has five impoundments. Three impoundments have impairment of beneficial uses. These impairments warrant priority in addressing problems and implementing acceptable cost-effective management practices.

Another basic concept to consider with reference to impoundments is that these structures not only serve for water storage but also act as treatment facilities (that is, large lagoons), which have a tendency to temper the degradation of water as it flows through a drainage. The settling of solids and utilization of nutrients result in better-quality water below an impoundment than above. However, the effectiveness and longevity of impoundments can be reduced when fed by tributaries that are laden with sediments. An example of this is Lake Powell on the Colorado River. Tons of natural salts and associated sediments enter Lake Powell each day.[6] Some estimate that the longevity of this impoundment will be reduced by 50 percent if sedimentation continues at the present rate. Control of sedimentation in the Colorado River system is primarily a function of land management on federal lands controlled by the National Forest Service and Bureau of Land Management. Several practices implemented by private landowners through federal and state sponsorship, such as contour management, could be implemented on federal lands but would be very expensive.

An example illustrating a controllable pollution problem is the Sevier

River System—an intrastate drainage system in Central Utah—where water allocations represent about 0.5 times the annual flow in the system. Strange as it may seem, this overallocation is possible due to irrigation return flow and groundwater seepage that returns to the system. However, irrigation return flow and groundwater seepage add significantly to the salt content of the water. The total dissolved solid content in the headwaters of the system increases at least fiftyfold by the time it reaches Delta, Utah, after passing though forests, meadows, agricultural land, and deserts.[7] In the water-poor Delta area, the water quality is marginal for irrigating alfalfa, a moderately salt-tolerant crop. Land management practices, such as sprinkler irrigation and prioritization of land use, should reduce the salt content in Sevier River water.

Education and Enforcement Problems

As is the case in other western states, water management officials in Utah have to grapple with two particularly tenacious problems: A general lack of understanding of water quality problems and issues; and violations of state standards. This is particularly true in rural communities where a lack of understanding of water quality problems results in a significant number of drinking water violations.

Efforts to educate, update, and inform local officials and citizens in communities throughout Utah on all facets of water pollution control, treatment, and importance are coordinated by the Division of Environmental Health. In this ongoing program, the division works closely with the thirteen local health departments in the state.

A unique and innovative program in Utah for identifying water violations is a computerized tracking system in the Bureau of Public Water Supplies. This system sorts through the tens of thousands of records for the nearly 1,300 public drinking water systems, flags violations, and sends a notice of violation to the responsible official with recommended action to correct the violation. The water systems are divided into two groups based on state regulations: community and noncommunity. A community water supply system serves at least fifteen service connections used on a year-round basis or regularly serves at least twenty-five year-round residents. A noncommunity water supply system is one that does not meet the community criteria.

According to recent data, over 66 percent of the bacteriological quality and the monitoring requirement violations occur in 64 percent of the community systems that supply less than 4 percent of the state's total population. Approximately one-third of the community violations occur with systems serving 96.9 percent of the population. Enforcement priorities logically meet the greatest need in the larger communities. Local health agency assistance in dealing with the small rural communities should result in a significant reduction of violations in rural communities. In dealing with violations in the noncommunity sector, emphasis will be placed on enforcement of state- and federal-agency-owned systems such as highway rest stops and recreation areas. Again local health agency assistance should reduce the violations in the noncommunity sector.

Other Water Quality Issues

Due to the highly mineralized nature of much of the state's landscape and surface crust through which aquifers flow, Utah is faced with a number of varying water quality problems. For instance, arsenic deposits in the Delta, Utah, area threaten an acceptable drinking water quality for some community wells. While a study has indicated that no significant health effects have resulted from higher-than-standard concentrations of arsenic in the Hinckley, Utah, water supply, an expansive project has been completed to solve the problems.

Mine waters in the Park City, Utah, area contain high concentrations of lead and arsenic. It has been proposed that some of these waters be used for public water supplies. The solution here is to dilute these mine waters with other "cleaner" sources to reduce the overall concentrations in the final product provided to the residents. In addition, the necessary water rights to supply this community growth must be negotiated.

In western and northeastern Utah, salts leached naturally into shallow aquifers result in a significant amount of shallow groundwater, which is not acceptable for culinary purposes based on standards. However, a number of residents have tapped and used these sources for decades. Many residents are so accustomed to this mineralized type of water that they consider the taste of pure spring water to be strange.

While Utah is an energy-rich state and possesses many uranium deposits, few groundwater or surface water sources exhibit radioactive levels that exceed federal guideline standards. Those waters that approach or exceed standards occur for the most part in the southeastern region of the state. Radioactivity does not seem to represent a significant problem for drinking water sources, at least at the present.

Technical Aspects of Water Treatment

During the past decade, Utah State University has undertaken studies to evaluate treatment technology and alternative treatment methodology to meet the future and particular needs of the state on a cost-effective basis. Studies have been conducted on lagoon systems, oxidation ditch treatment of sewage water, land application of secondary treated sewage, land filtration of treated sewage for nutrient removal, and disinfection techniques including chlorination, ultraviolet irradiation, and ozonation. Lagoon systems work effectively in rural southern Utah where inexpensive land is available and where water officials have limited knowledge of managing complex treatment systems and limited work force to maintain treatment systems. In metropolitan areas where land is at a premium and sewage quantities are large, more advanced and complex treatment systems are required to treat sewage. Polishing secondary treated sewage by sand filtration—to provide more efficient disinfection—is used in several treatment facilities prior to discharge. Polished secondary treated wastewater is Utah's goal for all municipal wastewater treatment facilities by 1985, although this date will probably shift to 1990 due to funding problems.

Oxidation processes to reduce sludge by aerobic digestion have been incorporated in several Utah treatment facilities. The oxidation process reduces the sewage treatment plant odors that are so objectionable to residents near the facilities. The oxidation ditch process, coupled with sand filtration and chlorination, provides discharge water that comes very close to culinary quality.

Land application of secondary treated sewage—to avoid costly nutrient removal—has been very successful where large populations are served in rural areas. This process has saved the public millions of dollars in construction costs, and additional savings in system maintenance. Land application appears to provide inexpensive removal and utilization of nutrients while providing a source of revenue for annual maintenance, when the water is used for irrigation of crops such as alfalfa.

The Bureau of Public Water Supplies has developed a pilot plant protocol and evaluation criteria. The protocol has enabled some communities to propose new and more cost-effective treatment technology.

Utah's Water Facility Needs

A summary of the estimated cost to provide needed wastewater facilities to serve the anticipated year 2000 population of the state is shown in table 13-2. It is anticipated that over two hundred projects must be initiated to meet this growth. Projects may consist of expansion of existing plants, improvements in treatment efficiency, new sewerline installation, and new treatment facility construction. The total estimated cost in 1985 dollars is $595.5 million for these facilities. Since most projects will not be initiated in 1985, a more realistic estimate, considering probable construction schedules

Table 13-2

Estimated Costs for Needed Wastewater Facilities in Utah to Serve the Year 2000 Population

(fiscal year projected cost in millions of dollars)[a]

	1982-1983	1983-1984	1984-1985	1985-1986	1986-1987	1987-2000
Cumulative state needs in 1982 dollars	179.0	351.9	400.4	496.5	535.3	595.5
Cumulative state needs in anticpated construction year dollars	179.8	366.0	422.8	544.5	597.8	765.4

[a]Based on the State Planning Coordinator's baseline year 2000 projection of 2,441,000 people.

and inflation, is approximately $765 million. These estimates are based on the state planning coordinator's baseline projection of 2,441,000 residents in the year 2000.[8]

Water treatment facility needs for public water supplies represent about $100 million. There are no federal grant funds to assist in this area. Funding occurs through bonding or revenue at the local level usually through municipal or district efforts.

Water Management and Infrastructure

The quality of Utah's water is regulated and managed by the Division of Environmental Health, Utah Department of Health. The Bureau of Water Pollution Control and Bureau of Public Water Supplies oversee the water quality programs of the state. Each bureau has a statutory regulatory committee appointed by the governor and receives a state appropriation, as well as federal grant allocations, to support program functions and planning.

Local Water Quality Management agencies were established under Section 208 of the federal Water Pollution Control Act Amendments of 1972 (FWPCAA). These agencies were instrumental in developing current local water quality plans for drainage basins and in determining the best management practices to meet local needs.

These agencies were primarily funded with federal 208 grants. Those funds have been drastically reduced with the expiration of the 208 time frame outlined in FWPCAA, 1972, and future funding sources for these agencies are not apparent. Many of these agencies and their functions will be absorbed by local associations of government or local health departments.

All water rights applications in Utah are administered by the Division of Water Rights, Department of Natural Resources. Water rights issues are legal aspects of the water quality that reduce or eliminate arguments concerning water ownership. This is extremely important in dealing with the allocation of interstate waters such as the Colorado River and its associated tributaries. There are seven states in the Colorado River drainage system, each vying for its appropriation of water to be placed in impoundments for various year-round uses. It is speculated that all the water in the Colorado River System will be appropriated by the 1990's.

Utah's share of the Colorado River System will be stored in the Central Utah Project, currently under construction. This system includes seven new reservoirs on the Green River System in eastern Utah's basin, and a transfer of Green River System water to western Utah and the Sevier River System, among others. Currently the use allocation in the Sevier River System is approximately 2.5 times the annual flow.

The development of water projects is administered through the Division of Water Resources, Department of Natural Resouces. Both the Division of Water Rights and the Division of Water Resources receive federal monies in addition to their state appropriations to fund their designated programs. The funding of the Central Utah Project has been limited and is most controversial at the federal level.

Funding: History, Problems, and Alternatives

Legal Aspects of Future Funding

The federal government has directed comparatively large sums of money into water projects in Utah where the state has played, for the most part, an administrative role in the federal program. As a consequence, state funding mechanisms for water pollution control projects have not been developed to any great extent. Local funding mechanisms have been developed only to the extent necessary to provide matching funds for federal grants.

In Utah, numerous state and local government entities have varied and diverse jurisdiction over water development and control. Table 13-3 shows a list of the state and local entities that have legal jurisdiction over various types of water-related projects and services.

There has been no state legal mechanism or entity developed to handle financing of water pollution projects. With the decrease in federal funding, the need for a legal mechanism with financing authority at the state level is evident.

Over the past year, the Utah Legislature has, through an interim committee, been holding hearings for the purpose of developing legislation to

Funding Clean Water in Utah 179

Table 13–3
Special Entities and Districts in Utah with Bonding and Taxation Authority

Entity	Utah Code Annotated	Purpose — Water Resources	Purpose — Water Treatment	Revenue Authority — Bonding[a]	Revenue Authority — Taxation
Special service and fees	11-23-4	X	X	GO bonds	Taxes and fees
Water conservancy district	73-9-3	X		GO bonds	Levy: 1 mill
Drainage districts	19-1-1.5	No longer exists		None	None
Metro water districts	73-8-3	X	X	GO and revenue bonds	Levy: 2.5 mill
Improvement districts	17-6-1		X	GO and revenue bonds	Levy: 4 mill
Irrigation districts	73-7-1	X		Bond elections	Levy and charges
Board of water resources	73-10-20	X		State bonding	None
County service area	17-29-3		X	Levy and GO bonds	Levy: 5 mill
Utah Safe Drinking Water Committee	26-12-3,5		X	None	None
Municipal Improvement districts	26-11-3			Revenue bonds	Levy: 1 mill
Water Pollution Control Committee			X	None	None
Soil and conservation districts	4-18-2	X		None	None
Interlocal Cooperation	11-13-2,3			Revenue bonds	In lieu tax
Public works	55-3-1		X	GO bonds	None

[a]GO bonds are general obligation bonds.

create a state solution to the funding problems. To date, draft legislation
has not been prepared, but the time frame for development requires introduction of a bill in the 1983 session of the legislature that began in January
1983. Alternatives being considered include a state grant program, a state
loan program, creation of a state bonding authority (debt management
assistance), and other state financial and technical assistance programs.

Wastewater Facility Funding

Since 1972 most major wastewater facility construction has been accomplished through the EPA Construction Grants Program. With the amendments to the Clean Water Act of 1981, the program was authorized at a
reduced funding level of $2.4 billion per year through FY 1985.[9] Figure 13–1
shows Utah's past funding appropriations and the expected funding
amounts through FY 1985. It is clear that funding levels will be reduced
over what has been available in the recent past due to federal budget cutbacks. Utah's share of the national appropriation will not exceed $11.8
million for FY 1982 and $12.8 million for FY 1983, 1984, and 1985. Fund-
ing beyond FY 1985 is uncertain at this time due to the New Federalism
proposals.

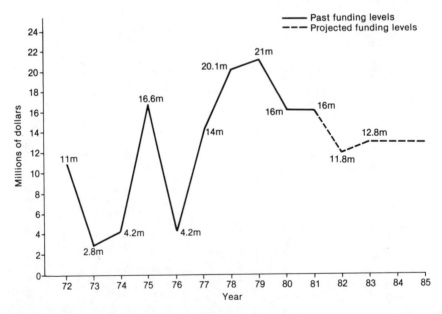

Figure 13–1. Construction Grants Funds Allocated to Utah under the
Federal Clean Water Act

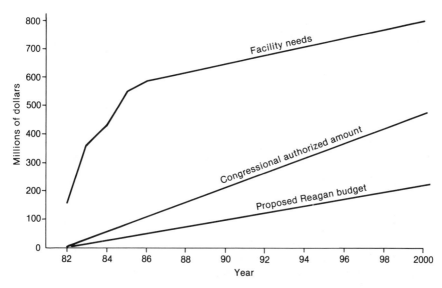

Figure 13-2. Cumulative Wastewater Facility Needs versus EPA Fund Commitments for Utah

Figure 13-2 compares projected EPA grant funds, assuming continued funding at the present authorized level, with the wastewater facility needs.[10] This figure illustrates the increasingly large shortfall of grant funds to meet state needs. The magnitude of this shortfall, and the fact that most communities have not budgeted to meet these needs without financial assistance, indicates the state is on the verge of a crisis in maintaining adequate wastewater facilities.

Figure 13-3 compares anticipated grant funding with estimated facility needs on an annual basis.[11] This figure illustrates that the needs and the fund shortfall are substantial in the first few years of the planning period. There is not time for lengthy rebudgeting plans and a smooth transition to solve the problem.

Other changes in the 1981 amendment affecting this program are: (1) The federal grant program will be reduced from 75 percent to 55 percent after October 1984; (2) only treatment plants and interceptor sewers to accommodate present population will be eligible for funding; and (3) federal assistance will no longer be allowed for growth of a community. These changes are most significant in Utah, especially with the population growth it will experience over the next twenty years.

Other funding sources, such as the State Community Impact Board, Community Development Block Grants, and the Farmers' Home Administration, have in the past not added more than 20 percent to the funds

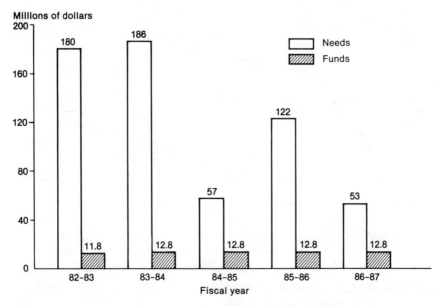

Figure 13-3. Total Need for All Projects Compared to Anticipated
Construction-Year Dollars

available through the EPA grant program. These funds, although very
helpful, will not significantly narrow the gap between needs and funds. It
is apparent that alternate funding sources must be developed to meet the
state needs.

Future Funding Options

Due to the financial difficulty of meeting the total state water facility needs
with traditional funding sources, a number of options will be considered.[12]

The first, simply stated, would be to do nothing. If growth is not re-
strained, this option would result in a tremendous increase in the pollutant
load to receiving streams, turning many of them into essentially open
sewers. This would constitute a totally unacceptable public health hazard
since these streams flow with unrestricted public access through the state.
Also the present beneficial uses of state waters would be greatly impaired or
lost altogether.

A second option is to restrict growth. The capacity and cost of new
facilities would be reduced if growth rates were restricted below present

planned projections. To artificially restrict growth would have serious political and economic consequences and would be difficult to implement. Many
public entities are relying on a growth rate to assist in bond repayment.

A relaxation of water quality standards is a third option. With the
reduction in federal funding, questions are raised about the need to provide
the levels of treatment presently mandated by effluent and water quality
standards. Obviously, there could be cost savings if treatment levels were
reduced. In fact, since Utah is an arid state with limited water resources, it
will be necessary to increase treatment levels as population grows just to
maintain the present quality of state waters. Since conventional treatment
processes only remove a percentage of the pollutants (usually 85 to 95 percent), the total pollutant load discharged will increase with population if the
level of treatment is not increased. Other pollution loads, such as urban
runoff and industrial spills, also increase with population growth. Unfortunately, the waste-assimilating capacity of receiving waters does not
increase.

As an example, table 13-4 shows the effects of increased population on
pollution loadings to the Jordan River from domestic wastewater treatment
plants. Removal efficiencies for various treatment levels are shown. It

Table 13-4
Analysis of Wastewater Discharge to Jordan River

Sewered Population	Percent Removal of Population Equivalent (BOD)[a]			
	85%	87.5%	92.5%	95%
50,000				
100,000	15,000	12,500	7,500	5,000
250,000	37,500	31,250	18,750	12,500
500,000	75,000	62,500	37,500	25,000
750,000	112,500	93,750	56,250	37,500
1,000,000	150,000	125,000	75,000	50,000
2,000,000	300,000	250,000	150,000	100,000

Source: Adapted from D.A. Ostler, "Meeting Utah's Future Wastewater Treatment Needs for
Health Protection and Water Quality" (Unpublished report, Utah Bureau of Water Pollution
Control, 1982).

[a]85.0 percent removal = EPA secondary treatment standards.
  87.5 percent removal = Utah secondary treatment standards.
  92.5 percent removal = Utah polished secondary standards.
  95.0 percent removal = Jordan River effluent standards.

At 100 gallons per capita per day:

  1,000,000 population = 100 mgd = 155 cfs
  2,000,000 population = 200 mgd = 309 cfs

A population equivalent = 0.17 pounds of BOD per day.

should be noted that with the 95 percent removal rate, the effluent standard for discharge to the Jordan River, and its population of 1 million, there is still an amount of pollution being discharged that equals the pollution effects of sewage flow from 50,000 people with no treatment at all. These data clearly indicate that the Jordan River effluent requirements will not result in a significant improvement to the stream but are needed just to maintain the conditions that existed approximately thirty years ago when 50,000 people were discharging untreated sewage to the Jordan River.

A fourth option would be to develop less costly treatment methods. Obviously, such would be ideal if methods could be developed to provide the necessary levels of wastewater treatment. The EPA has provided a 10 percent increase for innovative process design. Much emphasis is placed on the need for cost-effective wastewater treatment facilities.

A fifth option is to increase local funding. In order to finance needed facilities with reduced federal support, it will be necessary to increase local sewer rates to generate additional revenues. According to a recent survey, the average residential sewer service fee in Utah is $5.63 per month. These rates can be raised significantly and still remain within the realm of economic reality.

Water and sewer rates have been held at artificially low levels and as a result have been relatively unaffected by inflation (as compared to other public utility rates). Compared to the estimated cost of other public utilities such as telephone, electric power, and natural gas, these rates are a tremendous bargain for the service provided. If these rates were doubled or tripled, they would still be considerably less than other public utilities.

A sixth, and final, option would be to provide state fund assistance. There appears to be more and more acceptance of the need to return responsibility for constructing water facilities to the state and local level. Support is developing for a legislative proposal that is currently being prepared to provide state bonding assistance to local agencies. Proposed legislation would permit the state to sell revenue bonds and then contract repayment of these bonds with the sponsors of local projects. Such procedure should result in a better interest rate on the bonds and at the same time preserve local bonding capacities.

Conclusion

A high degree of health protection has been achieved in Utah through a vigorous program of constructing needed water treatment facilities. Since federal assistance will be diminished in the future, it will be more difficult to maintain this good record. It appears that necessary funds to construct needed water treatment facilities will have to come through increased local funding—and possibly some state assistance—if new legislation for a state bonding authority is passed. In fact, Utah and Missouri had the best bonding rating in the nation (Triple A) during 1981. State funding assistance through bonding legislation appears to be the most cost-effective basis to replace reduced federal funds. Equitable sewerage and service charges will assist in repayment of bonded loans.

Notes

1. D.A. Ostler, "Meeting Utah's Future Wastewater Treatment Needs for Health Protection and Water Quality" (unpublished report, Utah Bureau of Water Pollution Control, 1982).

2. Utah Bureau of Water Pollution Control, "State of Utah Clean Lake Inventory and Classification," draft report, 1982. U.S. Environmental Protection Agency, "Water Quality Assessment Summary, EPA Region VIII," 1981. U.S. Environmental Protection Agency, "Water Quality Trends, Region VIII," 1981.

3. Utah Bureau of Water Pollution Control, "State of Utah Clean Lake Inventory and Classification."

4. R. Loveless, "1982 Oral Status Report," Mountainlands Association of Governments, February 25, 1982.

5. M. Miner, "Oral Status Report: Weber River 208," February 25, 1982.

6. Southwestern Utah Association of Governments, "208 Water Quality Management Plan Update," 1981. Utah Bureau of Water Pollution Control, "Utah State Strategy for Salinity Control in the Colorado River Basin," 1981.

7. Six-County Commissioners Organization, "Sevier River Salinity Study: Summary Report," 1981.

8. Utah Office of State Planning Coordinator, "Utah: 2000—A High Development Scenario," 1980.

9. Ostler, "Meeting Utah's Future Wastewater Treatment Needs."

10. Ibid.

11. Ibid.

12. Ibid.

14 Funding Clean Water in Washington

Joan K. Thomas

Water Supply and Quality

The people of Washington State care about clean water. They have shown their care in public opinion surveys, by supporting strong state environmental laws, and, most important, by voting in favor of state and local bond issues for wastewater treatment facilities.

It is not surprising that the people of Washington have supported strong water pollution control laws and have been willing to spend tax dollars in achieving clean water goals. The generally excellent water quality of Washington State is the result of a combination of regulation and subsidized construction of wastewater treatment facilities, both public and private.

In 1967 the state legislature passed a law requiring the use of all known, available, and reasonable methods to prevent and control the pollution of the waters—both ground and surface—of the state.[1] In 1971 the legislature specified that "all known, available, and reasonable" treatment prior to discharge was required, *regardless* of the quality of the receiving water.[2] Long before the passage of the federal Water Pollution Control Amendments of 1972, now known as the Clean Water Act, the state of Washington required a state waste discharge permit for discharges to ground and surface water and for industrial discharges to municipal treatment plants.

The state's regulatory program was directed to industrial and municipal discharges, and it took many years to bring about compliance. In the mid-1950s Seattle had raw sewage outfalls on its downtown waterfront. Industries were providing little or no treatment to their process wastes, whether they discharged directly to ground or surface waters or into a city sewer. It became apparent to civic leaders and elected officials that the capacity of Washington's surface waters to assimilate waste had been reached and in many cases exceeded. It was also apparent that enforcement alone could not bring about the desired result. Pollution control means the expenditure of dollars that industry would have to justify to stockholders and that cities would have to justify to taxpayers. Thus the stick often is accompanied by a carrot.

The carrot in Washington's case came from state as well as federal sources. In 1967, the state legislature passed the industrial pollution control tax credit law.[3] Its purpose was to provide an incentive to industries to install pollution control devices and systems. It was intended that this tax credit program would last only two years, but, in fact, it was not terminated until 1982. Tax credits to date total $500 million.[4] The incentive to municipalities took the form of grants and loans. The voters of Washington have gone to the polls three times since 1968 to approve statewide general obligation bond issues for water pollution control. And the citizens of the Seattle metropolitan area voted in 1958 to tax themselves in order to clean up Lake Washington.

In 1968 the voters of the state approved Referendum 17, a $25 million bond issue for wastewater treatment. The sole purpose of Referendum 17 was to provide a 15 percent state share to those communities that received federal construction grants (a modest 30 percent of eligible cost in those days). On the same ballot were bond issues for outdoor recreation and juvenile institutional facilities. All passed. The $25 million for clean water was the largest dollar amount of the three, yet it received the most yes votes.

In 1972 voters were asked to approve a package of statewide general obligation bond issues that made up a program called "Jobs Now/Washington's Future." Of this package Referendum 26 was a $225-million bond issue for pollution control facilities; and, again, clean water carried the biggest price tag of the package, and, again, it led the ticket in favorable votes. Referendum 26 went beyond wastewater treatment and included money for solid waste, lake restoration, and agricultural pollution control. Referendum 26 was well timed in terms of providing a 15 percent state share to federally funded projects.

The 1972 Clean Water Act authorized 75 percent grants, and Congress appropriated $2 billion of the authorized $3 billion. In 1973, when the money started through the pipeline, many Washington communities were ready to construct their treatment plants. Thus in 1973 Washington was able to obligate nearly all of its $17.8 million federal grant money, matched with 15 percent state money, to treatment plant planning, design and construction—and this was the case until 1981. All a local community had to come up with was 10 percent of the eligible construction cost. A great many treatment plants in the state of Washington have been constructed with 90 percent grants.

In 1980 the voters were asked once more to provide a statewide general obligation bond issue, this time for $450 million, of which $315 million was earmarked for wastewater treatment facilities. Also included in Referendum 39 were $10 million for agricultural pollution control and $35 million for lake restoration. This time, water quality competed on the ballot with a bond issue for water supply, both domestic and agricultural, and water supply narrowly outpolled water quality but not by much.

With Referendum 39 the state of Washington embarked on a state grant program that is separate and distinct from the federal program. Proceeds of this bond issue are not used to provide a state share for federally funded projects. Instead, they are used for grants and loans of 50 percent of eligible costs as determined by the state and in accordance with a priority rating system.

The story of clean water in Washington would not be complete without discussion of the Municipality of Metropolitan Seattle (METRO). In the early 1950s residents who lived on or enjoyed the recreational and economic benefits of Lake Washington saw their metropolitan treasure degraded to the point that beneficial uses were lost or endangered.

While the quality of Lake Washington deteriorated, the citizenry was aroused by Seattle Attorney James R. Ellis, who called in 1953 for the creation of an advisory committee to "prepare a plan for the unified government of the metropolitan area and to go to the Legislature for the needed changes in state law." It took four years and two sessions of the legislature before legislation authorizing the formation of metropolitan municipal corporations was passed, and another five years before METRO was approved by the voters in 1958. Since then METRO has been—through its actions and accomplishments in the Seattle area, particularly in the clean-up of Lake Washington—an important force for clean water in Washington state.[5]

Washington State is blessed with another characteristic that is crucial to the protection of water quality: A clean water constituency. Citizen support is channeled through organizations such as the League of Women Voters and the Washington Environment Council (a statewide umbrella for local groups like the Audubon Society, Friends of the Earth, Sierra Club, garden clubs, sports fishing clubs, and neighborhood improvement groups). Civic organizations, business groups, and organized labor supported clean water. All states, of course, have such groups, but the support for clean water and the willingness to pay for it is not as great as in Washington. Moreover, Washington's constituency for clean water is well-distributed throughout the state.

The vicissitudes of federal funding for municipal wastewater construction grants since Public Law 92–500 was passed by Congress in 1972 are well known. The impact of this up-and-down pattern on Washington State allotments is shown in graph form in figure 14–1.

Until 1980 and the passage of Referendum 39, state bond proceeds were used primarily to provide state money to federally funded projects; thus the rate of construction was tied to the federal program. The state grant program funded by Referendum 39 is designed to fund more projects at a lesser share, and to return greater responsibility to wastewater treatment to the local level. The program has been well received, even in the depressed economy of Washington today.

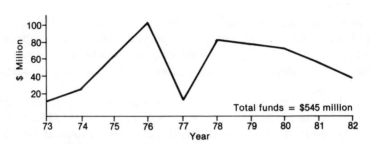

Washington State Department of Ecology, "Information on Sewer Facility Funding and Needs," unpublished paper presented to the State House of Representatives Committee on Local Government, Olympia, August 6, 1982.

Figure 14-1. The Level of Federal Grant Funding to Washington State Since 1973 under the Federal Water Pollution Control Act

Local, state, and federal agencies play very significant roles in Washington's clean water story. METRO is a prototype of the regional approach to wastewater collection and treatment; it is the wholesaler of wastewater treatment as it contracts with local agencies to treat their sewage and it performs areawide water quality planning and management. Other urban centers in the state have also implemented regional water quality programs and agencies.

At the state level, Washington was the first state in the nation to create a state agency to manage water quality, air quality, solid waste, shoreline protection, and water resources in an integrated and comprehensive way. The Department of Ecology was established by the legislature in 1970, and it brought together several similar agencies with narrower missions. Shoreline management was assigned to the Department of Ecology when the voters approved an initiative measure that established a planning and regulatory program to protect the fresh and saltwater shorelines of the state. During the late 1960s and early 1970s, a number of landmark environmental laws were enacted. The Water Resources Management Act of 1971 is significant because it recognizes the integral relationship between water quantity and water quality. The executive, the legislature, and the people have given the Department of Ecology a full set of tools to work with: regulatory authority, planning capability, management resources, and financial assistance programs. This combination has enabled the state to carry out a comprehensive environmental program that is both corrective and preventive.

The relationship between state government and federal regulatory agencies is strained at times, but on the whole it is characterized by cooperation and coordination. Region X of the U.S. Environmental Protection Agency is equally sensitive to the value people of Washington place on environmental quality and the willingness of state officials to administer federal regulatory and financial assistance programs. The construction grants program and the National Pollutant Discharge Elimination System (NPDES) permit program have been delegated to the state of Washington, and program assistance provided by Section 106 of the Clean Water Act has enabled the state to develop the necessary technical and management staff. This program grant may be endangered by the New Federalism, but at least for fiscal 1983, Congress has seen fit to appropriate slightly above the fiscal 1982 level.

Another chapter in Washington's clean water story relates the progress made in cleaning up industrial discharges. The forest products industry is very important to the economy of the Northwest, and the pulp and paper mills are a big part of the industry in Washington state. The sulfite pulping process uses enormous quantities of water and produces enormous quantities of wastes that can be discharged to air and water. Facilities to control and treat these wastes require enormous capital investments and entail high operating costs.

While problems still remain in this area, considerable progress has been made. Pulp mills in Washington have spent over $420 million on pollution control, of which more than $284 million was for water pollution. This has been offset somewhat by pollution control tax credits provided for in state law.[6] Thus far it has taken a combination of regulatory and financial assistance programs to upgrade municipal wastewater treatment, and the financial assistance came from the federal and state level. There is no state or federal grant program for industrial waste treatment; the present results have been achieved by regulation, enforcement, and tax incentives.

One final note on Washington's efforts to reduce water pollution involves the paper industry. Efforts to clean up the pulp mills started in 1962 with the Puget Sound Conference, which was called for by the governor under the terms of the federal Water Pollution Control Act. The conference involved Washington's Water Pollution Control Commission (now the Department of Ecology); the Federal Water Pollution Control Administration (now the EPA); and state and local health agencies. All sources of waste discharge into Puget Sound were examined, but the pulp mills received the most publicity. The conference ended in 1967 and enforcement efforts began. During the conference, scientific studies were conducted to evaluate the impact of pulp and paper mill wastes on Puget Sound's valuable fisheries resources.[7] Oysters, shellfish, and anadromous fish (salmon) were sampled and examined using state-of-the-art techniques. These studies showed that the four largest sulfite mills on Puget Sound discharged sulfite waste liquor in concentrations damaging to fisheries resources and that bottom sludge from solids in the discharge contained toxic components also harmful to fisheries resources.

As a result of the Puget Sound Conference, the four largest mills were required to design and construct facilities to remove 80 percent of spent sulfite liquor from mill effluent or limit (reduce) their discharges to a specified level by 1972. All mills were required to design and construct treatment facilities to remove all "settleable solids" from mill effluents prior to discharge by 1970, and six mills were required to dredge sludge beds from waterways to which they were discharged. All these requirements were to be enforced under state law, using the state's waste discharge permit authority (NPDES permits were not a requirement until the 1972 federal Clean Water Act).

There were to be sure many delays caused by legal challenges and some footdragging, but by 1975 some dramatic improvements in water quality could be seen. As the deadline for 80 percent recovery of waste sulfite liquor approached, the governor called for a state/federal/industry-financed study of the relationship between reductions in pulp mill waste discharges and biological changes in one embayment to which two mills in the city of Everett discharged.[8] From 1966–1971, total solids discharged into Everett Harbor routinely exceeded 2.5 million pounds per day. By the end of 1975 this had been reduced to less than 500,000. By December 1979 this was further reduced to 10,000 pounds per day. This was achieved by a combination of pollution control installation and reduced production. The biological impact of this was dramatic. The juvenile salmon livebox mortalities went from 100 percent to zero in all but one location. Port Gardner today is a popular place for fishing (both salmon and bottomfish), crabbing, and shrimping. Commercial oyster culture has returned to other parts of Puget Sound where pulp mills have either cleaned up their waste discharge or have closed down. The Olympia oyster, once an abundant delicacy in South Puget Sound, almost disappeared; now it is making a comeback.

Washington's achievements in water quality have come through the prodigious efforts of a caring citizenry, a concerned legislature, and conscientious administrators. The enormous cost in public and private dollars has been an investment in the quality of life and had brought an economic return by restoring and enhancing many of the beneficial uses of the waters of the State of Washington. To the people of Washington, clean water is essential to the kind of economic growth they want for their state.

In recent years scientific attention and public concern have turned more and more to toxic chemical pollution. Back in the 1950s technology could find parts per million; today it can detect parts per billion and, in some elements, even parts per trillion. If one looks at the water column, the cleanup has been a tremendous success; if one looks at the bottom sediments, it is a different story.[9]

Washington has its share of problems in this area. Commencement Bay, which is heavily industrialized and is located in the city of Tacoma, has been named by the Environmental Protection Agency as one of the potentially worst hazardous waste sites in the United States and has been made eligible for Superfund. All current waste dischargers, point and non-point in origin, have been sampled and analyzed for chemical pollutants, and it is clear that it is not current discharges that are causing the problem. Past discharges are not only difficult to trace, but their sediment-accumulated contaminants are extremely difficult and costly to correct.

North of Tacoma's Commencement Bay lies Seattle and its urban, industrial complex in the Duwamish River valley and its harbor in Elliott Bay. The Duwamish River is another toxic "hot spot," and Elliott Bay exhibits fish abnormalities and high concentrations of chemicals in sediments and fish organs. Although Everett Harbor is a success story in terms of the water column and current discharges, the National Oceanographic and Atmospheric Administration has found a higher incidence of bottom fish abnormalities in Everett Harbor than in either Commencement Bay or the Duwamish Waterway.[10]

Another knell of alarm has sounded: Toxic chemicals have been detected in the state's groundwater. City water supply wells have been closed in Tacoma and its suburb of Lakewood. Some of the groundwater contamination in Tacoma is thought to be hydrologically related to Commencement Bay, and the problem is being addressed through Superfund. The suburban wells, however, draw from the Chambers Creek/Clover Creek aquifer, a large and hydrogeologically complex system. The first step in trying to find the source of the chemicals and determine how they may migrate through the system is to map the aquifer.

Toxic chemical pollution is a problem that grows in magnitude as detection technology advances. Moreover, there are many more chemical compounds in use today, but knowledge of what happens to chemicals in the environment and of health risks has lagged behind the ability to create and use new chemicals. Many toxic pollutants are a legacy from the past—they will be found in sediments long after the point source discharge is corrected or eliminated, and they will be found in the groundwater long after the solid waste landfill has been cleaned up. Solutions to these water quality problems will be technically difficult and very expensive.

Meanwhile the problems caused by conventional pollutants have not been entirely solved. Nutrients, bacterial contamination, and low dissolved oxygen still degrade water quality in some areas of the state. A few examples follow.

The Spokane/Rathdrum aquifer has been designated as "sole source" under the federal Safe Drinking Water Act, and its protection is a high priority for local, state, and federal agencies. There is a high rate of exchange between the water of the Spokane River and the aquifer. On-site waste disposal systems and disposal of solid waste in landfills over the aquifer are a cause of concern, as is the treated domestic wastes that are discharged into the Spokane River by cities in Idaho and Washington.

The Yakima River, which flows east from the Cascade Mountains and into the Columbia River, has severe water quality problems in a stretch that runs through an area of intense agricultural and urban use. Irrigation return flows and treated domestic and food-processing wastes contribute to a problem that is aggravated by low flows during the irrigation season. The Yakima River once supported one of the largest anadromous fish runs in the Northwest. But diversion dams and flow depletion for irrigation have taken their toll, and the fisheries resouce has been almost destroyed. Most of the attention now being given to fish and wildlife restoration and enhancement is focused on water supply and management: adequate flows, fish passage facilities at dams; and competing uses (for example, fish flows versus irrigation). But fish need water quality as well as water quantity to thrive.

Shellfish are a valuable economic and recreational resource in Puget Sound and other estuaries; yet many commercial shellfish culture areas have been closed because of bacterial contamination. The eutrophication of fresh-water lakes is speeded up by nutrient loading and sedimentation. Lack of dissolved oxygen causes fish kills. Most of the lingering water quality problems in Washington can be attributed to non-point sources.

Urban run-off, agricultural run-off, animal waste, wind and water erosion, and on-site domestic waste disposal have all been addressed through Section 208 of the federal Clean Water Act. Progress in correction has been slow because the solutions are elusive. Non-point sources do not lend themselves to structural treatment facilities; they often require proper management of the activities that create the pollution. Dairy wastes can be stored in a lagoon for application to the land as fertilizer instead of being allowed to run off the land into the stream. Minimum tillage and proper management of irrigation water can reduce wind and water erosion and the consequent silting of streams. Responsibility for operation and maintenance of on-site waste disposal systems can be contracted to a public entity. Construction site erosion and sediment can be controlled by proper techniques. With few exceptions, however, the costs of correction are borne by individuals. State grants for agricultural pollution control and lake restoration are provided from Referendum 39, but the state constitution restricts eligibility to public bodies. There have been some creative approaches to combining state grants with federal cost-share and technical assistance programs in dairy waste management and lake restoration. The state's fifty-one conservation districts are involved in addressing agricultural pollution, and progress can be shown in many areas. Still, Washington has a long way to go in correcting non-point sources of pollution.

Water Management and Infrastructure

The Infrastructure

The economic well-being of any community depends on a good transportation system, an adequate and safe water supply, and the ability to dispose of its wastes. The above-ground and underground systems that perform these functions are part of the infrastructure that support civilization. The water infrastructure in Washington is, for the most part, publicly owned and operated. The capital outlay that built it came from tax dollars, through current levies or government borrowing; its operation and maintenance is generally financed by user charges or tax levies. This has been the norm in Washington State since it was a territory—and reflects its citizens strong populist tradition and belief.

Water Supply and Distribution

Most of Washington's public water systems are reasonably adequate, in a fair state of maintenance, and in generally good fiscal condition. The problems of repair and replacement, adequate and safe supply sources, and rate structures are not all comparable to those in New England and the Midwest. The physical system is relatively young (Washington became a state in 1889), and good mangement has kept it in good repair and allowed it to expand as needed.

It is noteworthy that there have not been large federal or state grant programs for water utilities; the water works industry has made a point of wanting to pay its own way. Most public water systems are run on a utility basis, and state regulatory authorities encourage rate structures that include capital recovery. There is a state grant program funded by statewide bond issues passed in 1972 and 1980. Grants are for no more than a third of eligible cost, and grantees must fulfill planning requirements in order to qualify. The planning requirement for water utilities is a ten-year plan with a five-year update, and it must include capital program, rate structure, and costs of operation and maintenance.

Since 1973 Washington State has committed $85 million for water supply improvement. Total construction since that time has been $250 million on over six hundred projects. There is an additional $100 million of identified work that needs to be done either to keep up with population growth or to maintain or replace equipment.

The administrator of the state water supply grant program is of the opinion that water utilities are well managed and in good fiscal condition because they never expected large federal or state grants. He also believes that the state grant program has been a success because it places most of the fiscal responsibility on the utility.[11]

Wastewater Collection and Treatment

Since 1964 Washington has spent $743.8 million in grant funds for wastewater treatment systems. Of the total spent, $566.3 million was federal and $177.5 million was state money. The local share of eligible cost was $176.9 million, plus all noneligible costs. Another $67.3 million in federal funds and $265.6 million in state funds will be obligated by July 1983.[12]

By far the greatest share of the total went to the planning, design, and construction of treatment plants and interceptors. Collector sewers generally have not been grant eligible, although some rehabilitation of existing collectors was funded.

The entire wastewater system in Washington is relatively new by eastern standards, but many of the underground pipes and trucks are much older than the treatment plant they supply. No estimate has been made of the cost of maintaining and replacing the present system. Every two years the EPA, with help from the states, prepares a *Needs Survey,* which is a cost estimate for construction of publicly owned wastewater treatment facilities needed to meet the requirements of the Clean Water Act. The following is the 1980 estimate of needs for the state of Washington.[13]

Category	Percent of Needs	Cost in Millions of Dollars
Treatment	29	907
Repair existing sewers	5	171
New collector sewers	16	499
New interceptor sewers	22	698
Combined sewer overflow	27	847
Total		3,122

The needs survey would appear to indicate that Washington's wastewater system is not adequate, but it should be remembered that the needs survey estimates what it will cost to meet the requirements of the Clean Water Act. There are systems in place that need upgrading of their treatment. They are not falling apart; they simply do not meet the letter of the law. The 1981 amendments to the Clean Water Act will save many small communities the capital and operating expense of building new treatment facilities. It has been recognized that some forms of treatment (for example, trickling filters, oxidation ponds, and lagoons) really do perform well, and the 1981 amendments call for a new definition of secondary treatment that will qualify these facilities. It is too soon to know how many Washington communities will qualify.

The state of maintenance of Washington's collection and treatment system is a mixed picture. In contrast to the water supply system, 90 percent grants have been available for wastewater since 1973. There is little incentive to maintain a plant when, for 10 percent of the cost, the city can build a new one. And once the plant is built, it may be too sophisticated for the town's unskilled and underpaid operator to run properly and to maintain in good working condition. This scenario is all too common for facilities built under Public Law 92-500 (1972 Clean Water Act) with additional grant assistance from the state.

Similarly, the generally poor fiscal condition of wastewater can be attributed to the obverse of the 90 percent grant; the 10 percent local share. Again there is little incentive to look closely at the cost of operating and maintaining this gift horse, much less replacing it. Raising rates is politically difficult; ratepayers would take it out on the town council at the next election. Wastewater systems in Washington are simply not run as utilities. Rates have been so low for so long that there surely would be a revolt if the city fathers tried to raise them to cover the true cost of service, including the use of assets.

Funding: History, Problems, and Alternatives

According to John F. Spencer, deputy director, Washington State Department of Ecology, "During the remainder of this decade, we will move toward more realistic objectives. We will continue to experience the tug and pull of a transition between the ambitious program of federal and state involvement and a new era of local self-sufficiency in the 1990's. In other words, user rates will go up."

In 1979 the highest monthly sewer charge in the state of Washington was $12.00; the lowest was $2.25. The average was $5.40 flat rate per month.[15] That year the average Seattle household paid $5.45 per month for sewer services, $2.53 per month for water, $5.60 per month for garbage collection, and $17.50 per month for electricity.

There is no doubt in my mind that by the 1990s public water and wastewater systems must be managed so that the major portion of the cost of expansion and replacement can be financed internally. If John Spencer's "new era of self-sufficiency in the 1990's" is to be achieved, then user rates will indeed go up. In Washington most water supply systems are much farther along the road to self-sufficiency than are the wastewater agencies. It will take many years to overcome the legal, institutional and political obstacles to local self-sufficiency. For the rest of the 1980s, the gulf between large federal/state subsidies and local responsibility must somehow be bridged. The inability of local agencies to generate the local share of a 50 percent state grant (and in many cases even the 25 percent required for a

75 percent federal grant) is the greatest problem in the state of Washington. Looking to the future through the 1980s, then, is to look at some interim financing concepts and some institutional rearrangements that will be needed if progress toward clean water is to continue.

The signals are not entirely clear, but I assume that the construction grant program will end in 1985; that the deadline for publicly owned treatment works to meet secondary treatment will remain 1988; and that the EPA will use enforcement in place of grants to achieve compliance. These assumptions are based on personal observation of congressional reaction to the administration's proposed amendments to the Clean Water Act, which is up for reauthorization in 1983, and on the experience of serving as a member of the EPA's management advisory group for four years ending in July 1982. The immediate future offers an opportunity to review the role played by local, state, and federal government in the 1960s and 1970s and to establish the appropriate role of each for the remainder of the 1980s and for the 1990s.

Legal Aspects

Should the Clean Water Act be relaxed in light of the expected 1985 cut-off in construction grant funding? It is clear that much of the progress toward clean water by municipalities has been tied to the availability of grant funds. If there is growing disenchantment with the notion that municipal waste disposal is a $90- to $100-billion federal responsibility, there is equal conviction that it cannot realistically become a $90- to $100-billion state and local responsibility either. To some, the answer may be to relax requirements or even to abandon the goals of the Clean Water Act. A more sensible approach would be to reconcile the nation's goals with available monetary resources, by agreeing on a uniform treatment standard to be attained and a priority process as a way to achieve it. This is the position the state of Washington has taken.

To abandon the goal would be a strong signal that municipalities are off the hook, and industry could be expected to ask why they should not be off the hook as well. Moreover, the investments made by both industry and government to date deserve better protection. States would be hard-pressed to maintain strong pollution control laws in the face of relaxed national requirements, and if state laws are relaxed, the inevitable day when little kids will once again stand wistfully on the beach in front of an "Unsafe for Bathing" sign (as they were depicted in the 1958 METRO campaign to clean up Lake Washington) will not be far behind. The cycle would begin again, but the cost of correction the second time around would be higher.

There has been a change of emphasis in administration of the Clean

Water Act; a change that gives more weight to water quality standards than to technology-based effluent limitations. It seems unlikely that Congress will back away from the National Pollutant Discharge Elimination System (NPDES) it established in 1972 when it also declared that it is a national goal to eliminate the discharge of pollutants by 1985. That goal probably will not be attained, but its declaration has served a useful purpose in rallying public support for clean water. The Clean Water Act should be given the rest of the 1980s to accomplish its purpose.

The state of Washington will continue to play a strong role in water pollution control under federal and state law. It has accepted delegation of the NPDES permit program and the construction grant program. In addition, there is the state grant program. The state's view of funding clean water in the future is well expressed in the speech made by the deputy director of the Department of Ecology to the Pacific Nortwest Pollution Control Association.

> Ultimately, a situation will be created where local government must contribute far more to the funding of waste disposal facilities; they will take more responsibility for how a project is carried out. Local government will become more accountable for all aspects of a project. Commensurate with this, however, should be a relaxation of the bureaucratic requirements in our present construction grant program.[16]

Technical Aspects

The 1980s will be looked back on as the decade when the struggle for clean water changed its target from conventional pollutants to toxics. Correcting discharge of biochemical oxygen demand in wasteloads was an engineering problem; correcting discharges of toxics is a scientific problem. A great deal of research needs to be done so that risks to human health can be determined and water quality criteria established. The technology for controlling toxic pollutants needs to catch up with the high-technology production that creates them. These needs are most appropriately filled at the federal level. States do not have the research capability to establish criteria; moreover, the criteria and standards should be uniform across the nation.

Another technical aspect that should be developed in the 1980s is innovative and alternative methods of wastewater treatment. In spite of the added incentives in the Clean Water Act for innovative or alternative treatment, the results have not met the expectations of Congress. It appears that the architects and engineers who act as consultants to local government in design of wastewater treatment facilities continue to rely on traditional processes and physical plants.

The remainder of the 1980s will see a shift in technology emphasis from "end of the pipe" controls to better management of the sources of toxic pollutants. Pretreatment of industrial waste is a controversial provision of the Clean Water Act; its pros and cons will be debated by Congress as it considers reauthorization. However, preliminary indications are that environmental organizations will resist any move to weaken pretreatment requirements.

A technical issue that has arisen in the state of Washington is whether existing treatment plants should be allowed to operate at less than current efficiency as affluent standards are relaxed. The state department of Ecology takes the position that backsliding is not permissible, and the department will continue to require that treatment facilities be operated in accordance with design efficiency.

Financial Aspects

Washington has had extraordinary success with statewide general obligation bond issues as a way of providing for state assistance in financing the capital costs of water supply and wastewater treatment facilities. The voters of Washington clearly support clean water. In the two years since Referendum 39 was approved and $450 million was authorized for wastewater, lake restoration, agricultural pollution, and solid waste facilities, the climate for bond financing has dramatically changed. There are two basic reasons for this: the state's depressed economy and the magnitude of the Washington Public Power Supply System's (WPPSS) bonded indebtedness. It was almost inconceivable in the fall of 1980 that current revenues would fall short of the amount needed to service the bonds or that the state's statutory debt ceiling, which is a percentage of current revenues, would be reached. Late last year the governor had to impose a moratorium on bond sales because of the debt ceiling.

The second factor, the WPPSS debt, may be even more significant in the long term. There is a possibility that the supply system will default on $7 billion in bonds sold to finance two partially built nuclear power plants that have been terminated. It is a long and complicated story, and the final resolution may take years in court, but the cloud of default hangs over the credit rating of every school district, city, county, and the state itself. In looking to the future through the 1980s, statewide general obligation bonds are not an answer in Washington.

The state's greatest problems in financing clean water have been identified as: (1) the inability of local governments to generate the local share of their project cost; (2) pressure on the state to fund collector sewers to eliminate public health problems; and (3) the need to either assure continued financial assistance to local governments or find a way to help them become self-sufficient in a reasonable length of time.[17] Recognition of these problems is shared by the executive and legislative branches of state government and by the statewide clean water constituency. Some of the avenues that are being explored will be outlined.

Washington is among the states that are looking very seriously at privatization. In September 1982 the state Department of Ecology and the Planning and Community Affairs Agency sponsored an informal workshop on the concept of privatization as proposed by Arthur Young and Company. As changes in tax laws have reduced the depreciation time for machinery and equipment to five years and for structures to fifteen years, the development of a wastewater treatment system can be an investment opportunity for the private sector. Private companies that have successfully financed solid waste/resource recovery facilities on a turnkey basis are said to be looking at the potential market in wastewater treatment. The Department of Ecology is reviewing the legal and institutional barriers to privatization as well as its feasibility for specific projects within the state. There are many questions about permits, enforcement standards, ratesetting, and the use of grant funds that will have to be answered, but the concept has many attractions to state and local officials hard-pressed to find the capital to build needed facilities.

Washington has been following closely some ideas being developed in New Jersey. That state commissioned Arthur Young and Company to study the concept of privatization; its governor has proposed the creation of the New Jersey Infrastructure and Financing Bank. The purpose of the infrastructure bank would be to assist local government units with loans or combinations of grants and loans to finance wastewater, water supply, resource recovery, and other purposes. The bank's capital would come from proceeds of general obligation bonds, federal construction grants for wastewater treatment, specific state revenues and, perhaps, private financing. Funds repaid, together with interest earned on loans, would provide a growing source of capital. Governor Thomas Kean has proposed to begin with $40 million of authorized but unobligated bond issue proceeds and the anticipated federal construct grant allocation. The proposal requires legislative action and approval by public referendum, and its fate remains to be seen. It is worth keeping an eye on.

The goal of the 1990s is local utility self-sufficiency. In one model for this (described in the June 1982 issue of *WATER/Engineering & Management*), user charges are based on a formula that includes cost of service, depreciation on the current value of assets (however acquired or financed), and a rate of return on the current value of capital employed during the year of operation. This generates an operating surplus out of which interest is paid. The remaining surplus is held (or, presumably, invested) for expansion and/or replacement of worn assets.[18] This model is so distant from current public practice in Washington that it would take many years to implement.

The state role in financing clean water continues to be an important one, but it is changing from that of grant dispenser to one that more resembles a financial consultant. Washington State for many years has had roving operators who assist communities in proper operation and maintenance of their treatment facilities; the 1980s may see the debut of roving financial experts who can provide assistance to local communities as they work their way toward self-sufficiency.

Notes

1. Chapter 90.48, Revised Code of Washington.
2. Chapter 90.52, Revised Code of Washington.
3. Chapter 82.34, Revised Code of Washington.
4. Donald Provost, special assistant for industrial and technical affairs, Washington State Department of Ecology, personal communication.
5. Municipality of Metropolitan Seattle, *The METRO Solution,* Technical Appendix No. 9 to Areawide Water Quality Plan for King County, Washington Cedar-Green River Basins (Seattle, February 1978).
6. Information courtesy of Donald Provost (personal communication).
7. U.S. Department of the Interior, Federal Water Pollution Control Administration; and State of Washington, Water Pollution Control Commission, *Pollutional Effects of Pulp and Paper Mill Wastes in Puget Sound* (Portland, Oregon, and Olympia, Wash.: 1967).
8. Washington State Department of Ecology, *Ecological Baseline and Monitoring Study for Port Gardner and Adjacent Waters; A Summary Report for the Years 1972 through 1975* (Olympia, Wash.: 1976).
9. See Donald C. Malins, et al., *Chemical Contaminants and Abnormalities in Fish and Invertebrates from Puget Sound,* NOAA Technical Memorandum OMPA-19. National Oceanographic and Atmospheric Administration/Office of Marine Pollution Assessment (Boulder, Colo.: 1982).
10. Donald C. Malins, NOAA, letter to Donald W. Moos, director, Washington State Department of Ecology, September 14, 1982.
11. Personal communication with Ken Merry, supervisor, Operations Unit, Water Supply and Waste Section, Office of Environmental Health Programs, Washington State Department of Social and Health Services.
12. Washington State Department of Ecology, "Municipal Construction Grants Obligations," memorandum from John Stetson, October 19, 1982.

13. Washington State Department of Ecology, "Information on Sewer Facility Financing."

14. John F. Spencer, deputy director, Washington State Department of Ecology, unpublished speech to the Pacific Nortwest Pollution Control Association, Portland, Oreg., October 22, 1981.

15. *1979 Municipal Utility Rate Survey of Washington State* (Seattle: Gardner Engineers, Inc., 1979).

16. John F. Spencer, unpublished speech.

17. Washington State Department of Ecology, "Information on Sewer Facility Financing."

18. E.J. Gilliland and Steve H. Henke, "Crisis: Financing Water and Wastewater," *WATER/Engineering and Management,* June 1982.

15 Funding Clean Water in Wisconsin

Paul N. Guthrie, Jr.

In Wisconsin water is an abundant resource. There are more than 7,563,442 acres of water within Wisconsin's borders, contained in more than 10,000 lakes and tens of thousands of miles of streams.[1]

From the beginning of the state's history, Wisconsin's well-being has been linked to its waters. The earliest European explorers followed the rivers, the lakes, the bays, and the portages that had been used for centuries before by native Americans. Along these travel corridors, new inhabitants settled and new economic activity developed. River junctions and portages were natural meeting points, and modern settlements grew where these early places of commerce were situated.

From this early recognition of water and its importance as a communications and travel medium and the English tradition of the crown's ownership of the tidelands and the beds of navigable waters[2] came the provision of the ordinance of 1787 which said in part:

> The navigable waters leading into the Mississippi and St. Lawrence, and the carrying places between the same, shall be common highways and forever free, as well as to the inhabitants of the said territory as to the citizens of the United States and those of any other states that may be admitted to the confederacy without any tax imposed or duty thereof.[3]

These concepts of the Northwest Ordinance were then passed along into the organic law of the territory of Wisconsin and finally locked into the fabric of modern Wisconsin with their adoption in the state constitution in 1848.[4]

To understand the pursuit of clean water in Wisconsin in the 1980s, one must understand these historical linkages. Water is Wisconsin's most important natural resource, and in its pursuit of pollution abatement, not only is the state fulfilling a need for the protection of the public health of its citizens but it is also carrying out its responsibility under what has become known as the "public trust doctrine."

All of this is not to say that the existence of the public trust doctrine—which has a long history of strong judicial sanction—has made the road to current pollution abatement an easy one.[5] It has not.

The modern understanding of public health needs has not come without much battle and great efforts by dedicated men and women. Throughout the history of Wisconsin, the road to water pollution abatement has been one of forcing municipalities and private entities to spend monies for the protection of their customers and their citizens. Generally, early state pollution statutes had an economic initiative rather than a public health and safety premise. Throughout Wisconsin's history, cost and private economics have retarded competent pollution abatement and appropriate infrastructural development.[6]

Having concluded that progress from statehood (1848) to the present has been slow and that public health has been secondary to economics, how can one explain Wisconsin's modern successes? The answer is much too complicated to be fully explored here, but there are several critical elements involved that can be discussed.

First, the nineteenth century was a century of local responsibility with limited communications. Even when state legislation seemed comprehensive, it often left issues totally to local administration and thus vulnerable to local manipulation. In this century and especially in Wisconsin, with the advent of progressive state government, an overall statewide consistency evolved.

Second, while the battle for public health in water proceeded at a frustrating pace through decades of indecision and battle, a more subtle force began to take shape, anchored in the trust doctrine and articulated in judicial opinion. And while the concept may have begun with issues of navigation, commerce, and access, by the time of *Just* v. *Marinette* (56 Wisc. 2nd 7), a judge would declare, "The State of Wisconsin under the Trust Doctrine has the duty to eradicate the present pollution and to prevent further pollution in its navigable waters."

The subsequent joining of the public health concern with the public trust doctrine has generated the current public support for abatement. This basic support is critical and fundamental to a statewide effort at water pollution abatement, whether financed by local dollars or state dollars. And until wastewater pollution abatement has been accepted as a fundamental public responsibility in the same degree of necessity as police and fire protection, garbage pickup or snowplowing, no amount of exhortation of worthiness will provide the basis for adequate financing.

In Wisconsin this degree of public acceptance is, and has been, large. For example, in a 1978 poll produced by the Wisconsin Center for Public Policy, a carefully selected cross-section of adult Wisconsin residents, indicated that environmental problems were second only to taxes as the state's biggest problem.[7] And when asked to identify (on a scale of 1-7), "How much do you worry about environmental problems?" the following resulted:

Never	*1*	*2*	*3*	*4*	*5*	*6*	*7*	*Very Often*
	7%	9%	12%	19%	18%	15%	20%	

In a 1981 survey it was reported that "the Wisconsin adult population is very interested in nature and the environment." Of all persons interviewed, 81 percent were very interested in at least one nature topic, and 87 percent were very interested in at least one environmental quality/management topic.

Findings in recent national polls suggest that national opinion is quite similar.[9] It is likely that a full focusing of this basic public support can be achieved only after current economic issues are resolved. In those states where the same degree of the merging of public initiatives has not occurred, the task will be harder.

Grants to Local Government

State Funding: A History

The first Wisconsin state water pollution abatement grant program originated in August 1966. This program was created to provide 25 percent supplementary grants—as required by the federal Water Pollution Control Act (Public Law 84-660)—to enable communities to receive federal (50–55 percent) grants. Before 1970 the state made forty grant offers, and the state legislature annually appropriated only enough money to make the necessary grant payments.

In January 1970 a new program titled the ORAP-200 (Outdoor Resources Action Program) began allocating to the Department of Natural Resources $100 million over a ten-year period to assist municipal water pollution control facility development. An additional $13,085,000 was added in 1977. Initially this program provided 25 percent grants to supplement federal grants, and 25 percent grants for pollution control projects constructed after August 1, 1966, that received no federal assistance.

After the enactment of Public Law 92-500 (October 18, 1972, Clean Water Act), the state program was divided into 5 percent, 15 percent, 25 percent, and 50 percent cost-sharing possibility. The 5 percent share was a supplemental grant to 75 percent federal grants provided under Public Law 92-500. The 15 percent share was also made in addition to U.S. Environmental Protection Agency 75 percent grants and was available only to those communities that had to provide a higher level of treatment than secondary treatment and only for the incremental portion of the cost attributable to advance treatment. Twenty-five percent grants were available for projects not funded by EPA 75 percent grants and projects such as sanitary sewer collection systems and combined sewer separation projects, sewage treatment facilities, and interceptors. Over six hundred 25 percent grants were made.

The 50 percent program was adopted to provide an alternative for economically depressed, unsewered communities that were low on federal priority lists. From the $144 million in the ORAP-200, $5 million was reserved for this purpose, and fifteen projects were funded.

Federal Funding: A History

Between 1955 and 1966 there was a federal grant program that provided a maximum grant of 30 percent. However, each year a dollar limit that could be given each project was set. During this period 225 federal grant offers were made totaling $51,904,183.

Between 1966 and 1972 it was technically possible for a community to receive a maximum grant of 55 percent from the federal government. However, due to the inadequate appropriations by Congress and the large cost of some projects, most communities initially received less than 55 percent maximum. In addition, a community was eligible for a 10 percent grant increase if they participated in regional planning for joint treatment. During the 1966–1972 period, there were 89 projects (construction costs: $181,834,174) that were funded at the 50 percent level or less, and 166 projects (construction costs: $101,875,000) that received a maximum of 55 percent. Since 1972, under Public Law 92–500, the basic federal grant share has been 75 percent of eligible cost. To date, federal grants totaling $581,017,970 have been awarded to Wisconsin communities.[10]

Current State Programs

As ORAP began to wind down and the EPA program continued to be mired in its own web of administrative ineptitude and congressional inaction, 1977 became a turning point for pollution abatement in Wisconsin.

As it became clear in late 1977 that the goals of the Clean Water Act were to remain intact and that existing levels of financing (ORAP plus federal) were not adequate to meet national objectives, the Wisconsin legislature began to grapple with how to achieve compliance with the federal act within the time limits prescribed by law. In this introspection certain conclusions became evident:

The lack of timely and consistent federal grant allocations destroyed the ability to expeditiously plan and build needed projects.

A shortage of planned and designed projects ready for construction caused unnecessary construction delay when money was made available.

Given the needs, and rapidly accelerating inflation, the federal grant sums were not sufficient to carry out the federal law, and, in fact, because of inflation, ground was being lost.

A single system of reviews, irrespective of funding source and requirements, for all municipal wastewater projects was essential to prompt project completion.

To promote expeditious and sincere community action, a single priority system, equitably designed, evenly administered, and essentially stabled in its factors for an extended period of years, was necessary to provide local governments with a sense of progress and anticipation of financial success.

Uniform statewide enforcement of Clean Water Act violations would need to be maintained to encourage timely completion of projects.

From this concern came the Wisconsin Fund,[11] which was designed to try to satisfy many of the shortcomings of prior activities and to make possible a coordinated management of construction. The fund was designed to: hold a management system together; provide adequate resources in a timely manner; making planning and design adhere to a time schedule by having monies available on time; consolidate review staffs from both the federal and the state programs with the assistance of delegation under the Cleveland-Wright amendment; and coordinate planning and design with consistent and applicable priority setting.[12]

The Wisconsin Fund's philosophical premise was simple: that because of the vicissitudes of the federal program, no soundly conceived long-term project scheduling could be achieved and, as a result, costs were uncontrollable. As a grant program, the fund's enabling statute is simple. It provides for a system of community grants similar in nature to the federal grant program, and although some differences in eligible costs were enacted, basic project concepts were maintained. Basic conditions such as the adoption of user-fee systems and operation and maintenance plans were required, and the federal priority system was adopted. Grant shares of 60 percent for Step 3 and 75 percent for Steps 1 and 2 grants were adopted. Since May 1978, 248 grants have been awarded, totaling $265,603,369. In addition, another $147,168,270 has been committed for spring 1983 construction. (For an explanation of the "steps" see page 99).

Current Needs and Results

In 1972 Congress rewrote the water pollution laws of the United States, declaring: "The objective of this act is to restore and maintain the chemical physical and biological integrity of the nation's waters . . .:

1. It is the national goal that the discharge of pollutants into the navigable waters be eliminated by 1985.
2. It is the national goal that whenever attainable, an interim goal of water

quality which provides for the protection and propagation of fish,
shellfish and wildlife and provides for recreation in and on the water
be achieved by July 1, 1983."[13]

Under the Clean Water Act, municipal treatment plants are required to
achieve certain water quality objectives. Estimates of total expenditures
necessary to reach these levels have been difficult to develop, and in the
early days of the act, they were very bad. However, over time a basic system
of biennial cost estimation developed by the EPA has begun to serve as a
good vehicle for decision makers.

This system—although hampered on the national analysis level by
political decisions changing the need definitions to fit particular national
government budgetary purposes—has served Wisconsin well because of the
state's decision to complete the 1976 survey on a 100 percent on-site survey
basis.

Each subsequent survey has been designed as a correction from that
base, modified to reflect changing conditions, court decisions, and final
cost estimates. In 1977 it was estimated that $2.3 billion in 1976 dollar costs
would be necessary to meet the pollution standards between 1977 and 1990.
It was also estimated that these costs would grow by $324 million during the
same thirteen years because of increased treatment need and capacity
generated by population growth. Current estimates, based on 1982 data,
show for the first time an impact from the Wisconsin Fund expenditure and
a *real* lessening of need.

Current Estimated Needs

In viewing future fiscal impacts of pollution abatement in Wisconsin, it is
necessary to review current needs estimates. To fully interpret the fiscal
situation for Wisconsin communities, it is necessary to separate needs into
two categories: the Milwaukee metropolitan area project and the balance
of the state. This is done for the needs data as presented in tables 15-1
and 15-2.

To simplify analysis in this discussion, both backlog and design cost for
year 2000 numbers will be reviewed—namely, that construction necessary to
become current in facilities for existing or proposed permits and that
construction necessary to meet year 2000 needs. In actual facts, few communities can build only to current needs levels. Most new tretment plants
are designed and constructed to accommodate need over the life of the
facility. This reserve capacity for treatment facilities usually reflects a
design life of approximately twenty years (thus, the year 2000). In the
1982 needs data for treatment plants, the buildable cost is estimated at

Table 15-1
Wisconsin Municipal Needs, 1982
(thousands of dollars)

Project Category	Milwaukee[a]		Balance of State	
	Backlog	Year 2000	Backlog	Year 2000
Secondary treatment	564,833	564,883	220,000	331,000
Advanced secondary			86,000	143,000
Advanced treatment Inflow/infiltration	120	120	32,000	32,000
Major sewer rehabilitation	71,265	71,265	7,735	8,735
New collectors	21,712	21,712	188,288	266,288
New interceptors	406,262	406,262	139,738	213,738
Combined sewer overflow	243,195	243,195	125,829	125,829
Totals (000s)	$1,307,437	1,307,437	799,590 — 2,107,027	1,120,590 — 2,428,027

Source: Adapted from U.S. Environmental Protection Agency, Office of Water Program Operations, *1982 Needs Survey, Cost Estimates for Construction of Publicly-Owned Utilities Treatment Facilities.*
[a]Milwaukee costs are court ordered, and no reserve factor for year 2000 is calculated.

Table 15-2
Wisconsin Municipal Backlog Needs, FY 1984
(dollars)

	Milwaukee	Balance of State
12/31/81 Backlog needs	1,307,437	799,590
1981 construction	− 78,138	− 73,680
1982–83 fundings	− 143,000	− 102,000
Net balance FY 1984 backlog	1,086,299	623,910

Source: Adapted from U.S. Environmental Protection Agency, Office of Water Program Operations, *1982 Needs Survey, Cost Estimates for Construction of Publicly-Owned Utilities Treatment Facilities.*

$1,038,000,000, as opposed to the backlog cost of $870,000,000. Grant assistance programs do not recognize these reserve capacities for full funding, and local communities must bear much of the differences in costs.

Balance of State: Analysis: As the preceding tables indicate, by the end of the 1983 construction year, outstate Wisconsin (that part of the state outside of the Milwaukee metropolitan area) will have approximately $623,910,000 in unconstructed backlog projects. Of this amount, $188,288,000 were identified as needed new collector systems—elements that are not generally necessary for immediate compliance with the Clean Water Act. The data further identifies some need for combined sewer work ($125,829,000) and a significant amout of new interceptor activity. Much of this latter category would be in conjunction with new treatment facilities under construction or planned.

Milwaukee Metropolitan Area: Analysis: For the Milwaukee area, the situation is more pressing. As of December 31, 1981, unmet Milwaukee project needs are estimated at $1,307,437,000. At the end of 1983, it is further estimated that $1,086,299,000 will remain. Under court orders currently in place, the Milwaukee metropolitan district is ordered to do the following work.[14]

Year	Construction Cash Flow (thousands of dollars)
1984	194,000
1985	195,000
1986	166,000
1987	141,000
1988–1995	586,000

In terms of the types of construction, Milwaukee represents the classic larger city. Treatment facilities in need of major work consist of 43 percent of the total cost. Major sewer rehabilitation (5 percent) is less extensive than one might expect, but new interceptors (31 percent), largely to accommodate already existing and overflowing volumes and to interconnnect the two treatment facilities, are a major cost. Finally, correction of ancient and overflowing combined sewer/stormwater systems of the older city is estimated at 18 percent.

General Analysis

Overall, in fiscal terms, Wisconsin's most serious problem in project development financing will be in the Milwaukee project. By and large, by the middle of this decade outstate communities will have completed their new construction, and their fiscal crisis will be in the cost of operation and maintenance, retiring acquired debt, and recovery of depreciation. In Milwaukee, construction must continue for the next thirteen to fifteen years, and financing the costs for this construction will extend even longer.

Current Results

Because of the massive needs when the current federal program was adopted in 1972, the huge underestimation of cost, the relative small amount of federal funding, and the long length of time necessary to complete a project, visible water quality improvement has been slow in appearing to the average citizen. As a result, the political process of obtaining continued funding has been seriously affected. Yet, contrary to discouraging reports from some financial/political soothsayers, the past decade's expenditures and water enforcement policy are beginning to show strong positive results in Wisconsin. And as a result, public support and funding has continued.

For example, a comparison of 1977–1978 data with 1980–1981 data from the statewide water quality ambient monitoring network shows that during this period, total phosphorus and suspended solids (SS) was reduced approximately one-third, and the average biochemical oxygen demand (BOD) decreased at 75 percent of the sampling locations.[15] Such data is a mark of a decade of municipal and industrial pollution abatement efforts. Specific pre- and postoperational studies of facilities constructed in recent years document discernible water quality improvement that is directly attributable to improved treatment.

In terms of compliance with state and federal law, Wisconsin industry and municipalities are proceeding effectively. As of June 1981, there are 568 municipal treatment facilities discharging in the waters of the state. Of this number, 73 (13 percent) utilize land disposal; 281 (49 percent) are required to treat to secondary treatment levels; 50 (30 mg/1 BOD and 30 mg/1 SS) have limits of 15 mg/1 BOD/20 mg/1 SS; 87 (15 percent) are required to meet 20/20 BOD/SS; and 77 (14 percent) have limits defined by allocation based on stream conditions.

Since 1978 over a hundred municipal systems have been upgraded, and by 1983, more than 120 more will have been completed. In the most recent reporting period, 95 percent of all newly constructed municipal treatment works were meeting the legal discharge requirements. By 1983–1984 it is estimated that approximately 85–90 percent of all municipal facilities will be meeting legal discharge levels. Most Wisconsin industrial dischargers have met "best practicable control technology currently available" (BPCT), thus providing major improvement for the state's waters. For example, in the decade of the 1970s, discharge limits established for the state's forty-seven pulp and paper mills resulted in a 90 percent reduction in BOD and a 75 percent reduction in SS, even with large increases in production.[16] Other industrial groupings show similar compliance.

In the summer of 1982, 95 percent of all major dischargers were meeting permit requirements. Recently a total of 123 municipalities were under enforcement obligation to construct or renovate treatment facilities. Exclusive of the Milwaukee Metropolitan Project, these communities represent a potential cost of $200,077,654.

Issues and Problems

One of the issues that cannot be overlooked in any analysis of long-term wastewater management policy, both in Wisconsin and elsewhere, is the structure of the serving organizations.

Structure

In Wisconsin there are at least nine distinct types of statutorily approved municipal units of government providing sewer services. These include: joint sewerage districts (pursuant to Section 144.07, Wis. Stats.); metropolitan sewerage districts (Section 66.04); city and village governments (Section 62.18); town sanitary districts (Section 66.072 and 60.30); town utility districts (Section 66.027); county utility districts (Section 59.083); intergovernmental contract districts (Section 66.30); and the Milwaukee metropolitan district (Chapter 282, Laws of 1981). Each of these operate somewhat differently, and each type has a unique organic enabling statute. And, although each of these operating systems is appropriately permitted and regulated for pollutant discharge, the lack of consistency and structure among them does cause problems. This is especially true in terms of management and fiscal capacity. As treatment facilities and systems are constructed and become more sophisticated and more costly to build, operate, and maintain, many of these districts will become hard-pressed to function effectively and efficiently.

In practical terms, a legally organized district is viewed in today's managerial and fiscal world as rebuttably presumed to have a the capacity to carry out its chartered activities. In many instances the elimination of this presumption will occur only at the point of crisis or nonperformance. To some, such a risk of large-scale public capital investments with inadequate operating entities is an item of concern, but when coupled with the obvious environmental risk associated with nonperformance, it is critical.

In Wisconsin, and possibly in many other states, the structure of managing agencies needs serious attention. Such a concern is further highlighted by the increased need to install adequate operation and maintenance practices in recently completed facilities. Fundamental to this effort is the need to establish local community wastewater treatment as an "important" public responsibility. No longer is it possible for the small community to staff its treatment facilities with the unused time of street workers, snowplow operators, or park employees. Modern facilities require trained personnel to manage and regulate operations.

Because of the large costs of new treatment facilities, capital cost acquisition and management very understandably has received the primary attention in wastewater abatement. Nevertheless, the operation of wastewater treatment services is now one of the biggest costs of municipal government; and wastewater management also has become important. Municipal finance data for Wisconsin cities, villages, towns, and counties indicate annual expenditures of approximately $150 million for operation. A comparable amount was expended on capital projects.

Utility Governance

A new element has emerged in recent years in the fiscal arena. Historically wastewater treatment has been a minor element in a typical user's cost. Rates have been established on an indifferent basis with little or no oversight, and depreciation accounts for plant and equipment have either not existed or have been used for shoring up other aspects of municipal budgets. With the passage of Public Law 92–500, Congress required the establishment of user-charge systems to protect the fiscal integrity of funded projects.

In Wisconsin this concern has been emulated in the Wisconsin Fund. However, this fiscal efficiency concern has not as yet passed from the granting sections of the law and regulations to the governing statutes. And although dischargers must perform adequately, except under grant statutes, only in two circumstances are rates and charges subject to public agency oversight at the state level: rates of joint water-sewer agencies and an equity appeal to the Public Service Commission.

Such a condition is both surprising and worrisome because as construction costs have increased and operations become more expensive, the public cost has become substantial. Wastewater management is clearly a public utility within the philosophic context that has led to utility oversight of other community systems such as electric power, telephone, and drinking

water. Yet just a few of these public protections now exist. There are
perhaps three reasons for this. First, as long as the basic grant programs
continue to operate and to some measure provide oversight of user systems
additional oversight allegedly may be unnecessary. Second, the fundamental tie between environmental performance requirements and costs is so
important that any attempt to regulate costs and charges may infringe upon
environmental mandates. Since in Wisconsin and many other states rate
setting and environmental regulation are managed by different state agencies, the generic conflict is observed. Third, the county is in an era of anti-
regulation. The historical progressive tradition that government has an
important fundamental function in the protection of its citizens and adjudi-
cating disputes between contending interests has to some extent been lost in
the rhetoric of deregulation and decentralization.

However, this concern will not go away, and whether a state enforces
the concepts of adequate fiscal and managerial protection through the audit
process (with mandatory periodic audits), through the rate-setting process
(similar to public utility control), or through environmental regulation and
permitting (an extension of the permitting allowing pollution discharge), it
must be addressed. In this context, two points are clear: (1) The local competitive political process will not do the job, as history has shown, and (2)
ignoring the problem will not cause it to dissipate. Over the long haul, the
effectiveness and full utilization of recently built and to-be-built systems
requires careful review of the local managerial structure and of government
oversight.

Milwaukee: The Large City Problem

Overwhelming in the data on needs and costs are the large anticipated costs
of the Milwaukee metropolitan area. The 1982 needs and enforcement data
confirm that by the 1985 funding year, most outstate Wisconsin enforcement requirements projects will have been started. In Milwaukee, construction will take much longer, and, indeed, the managing and financing of this
activity is the single most significant wastewater issue now facing the state.

To date, the project has moved near to schedule with grant assistance
available for most eligible elements. Approximately $100 million has been
spent in bringing comprehensive, systemwide planning to the design stage.
Currently, design is underway on most elements, and the first stages of
construction are beginning.

In the next six years, cash-flow needs of the district may range in the
magnitude of $200–$250 million per year. Grants from state and federal
sources could average $90–$100 million, thus requiring $100–$150 in annual
local dollar commitments. This is for a jurisdiction that until recently

(1970–1975) spent only about $15–$20 million a year on wastewater projects. After lengthy litigation, in 1977 and reaffirmed and modified in September 1982, by the Dane County Circuit Court, the Milwaukee Metropolitan Sewerage District is under order to rebuild its system at an estimated cost of $1.6 billion.

It is expected that local-share acquisition will involve a variety of financing tools, including general property tax levies (under recent legislation now available to the newly reconstructed commission), general obligation bonds, system revenues, and perhaps revenue bonds of the district. Efforts at increasing state assistance will continue. At present, approximately $50 million a year is available to the district. Future efforts will undoubtedly evolve around increased state support, lengthened construction schedules, and state funding for bonds.

In this era of financial stress, it is becoming more and more difficult to develop comprehensive financing flows, and therefore, to maintain construction scheduling, and thereby minimize costs. The implications of this dilemma are more than simply a Milwaukee problem:

If Milwaukee does not follow its orders, why should anyone else in the state comply with the law?

A failure to complete facilities would weaken the Milwaukee area in economic competition for new employment because it lacks the "infrastructure" to accept new major discharges.

If not carefully constructed, a financing plan for Milwaukee could cause economic distress to the region by overloading the tax and bond structures, thus driving out employees and investors.

Milwaukee is a classic big city problem—and is, perhaps, a prototype for other communities.

Policy and Regulation

A final area of concern for the future is the direction of federal policy. Fundamental to the success of abatement efforts in Wisconsin has been stable priorities, stable plans, stable financing, and stable policy and statutory enforcement. With these ingredients, communities can work to meet their wastewater needs. Without this stability, local leaders are unable to muster political support, or to identify adequate finances to complete projects.

In the current climate of national environmental uncertainty (even if 80 percent of the public disagrees), with the current administration's environmental prescriptions, some local communities will wish to wait out the process, hoping that pollution abatement will go away. It won't, and in the end, it will only cost more.

But in an era when a national administration is trying to test the climates for a return to case-by-case, deal-by-deal water quality control, and uniform national standards (so critical for balanced national economic impacts of environmental regulations) are under attack, the problem for administrators (who plan, design, and schedule for the future) is: Where are we going?

Notes

1. Wisconsin Department of Natural Resources, "Economic Impact Statement," Wisconsin State Grant Program Budget Request," November 1976, p. 13.

2. Victor J. Tannacone, et al., *Environmental Rights and Remedies,* I (Rochester, N.Y.: Lawyers Cooperative Publishing Company, 1977), p. 16.

3. U.S. Stats. 51, Article 4, an Ordinance for the Government of the Territory of the United States North West of the River Ohio, July 13, 1787.

4. U.S. Stats., Chapter LIV, Section 12, April 20, 1836, Article IX, Section I, Constitution of the State of Wisconsin, adopted March 3, 1848.

5. See *Dana Shooting club* v. *Hustings,* 156 Wisconsin 261, 271; *In Re Crawford, L&D District,* 18 Wisconsin 409, *Just* v. *Marinetta County,* 56 Wisconsin 2nd 7.

6. See Earl Finbar Murphy, *Water Purity, A Study in Legal Control of Natural Resources of Wisconsin* (Madison: University of Wisconsin Press, 1961), and *The Natural Resources of Wisconsin* (Madison: 1956).

7. *The People on Wisconsin,* (Madison: Wisconsin Center for Public Policy, 1978). A similar finding is found in *Public Survey,* Wisconsin Department of Natural Resources Office of Policy and Analysis, conducted by the University of Wisconsin Survey Research Laboratory, and *A Survey of the Attitudes of Wisconsin Citizens Towards Public Policy and their Media,* August 1976. Wisconsin Center for Public Policy Studies, Hart Research Associates, Inc.

8. *Environmental Education Needs and Interest of Wisconsin Adults* (Madison: University of Wisconsin, 1981), p. 61.

9. See Louis Harris, "Public United in Concerned about the Environment," *Common Cause,* June 1982, p. 6.. Also see "The Harris Poll Predicts no Tilt to the Right in November," *Milwaukee Journal,* August 12, 1982.

10. Wisconsin Department of Natural Resources, *Status of State Wastewater Grant Programs,* unpublished paper, July 18, 1975.

11. 44.24 Wis. Stats.

12. Section 205(g)(2), Federal Clean Water Act.

13. Public Law 92-500, Section 101(a).

14. Milwaukee Metropolitan Sewerage Commission, "Cash Needs 1983 Capital Budget," *Capital Improvement Program* (Milwaukee, Wisc.: 1982).

15. Wisconsin Department of Natural Resources, *Wisconsin Water Quality 1982,* Report to Congress, April 1982, pp. 10–13.

16. Ibid., p. 36.

Tax Policy
Roundtable Members

Roy W. Bahl
Syracuse University
Syracuse, New York

Marion S. Beaumont
California State University
  at Long Beach
Long Beach, California

Kenneth Back
Property Tax Institute
Washington, D.C.

Charles D. Cook
Lincoln Institute of Land Policy
Cambridge, Massachusetts

Harvey B. Gantt
Gantt-Haberman Associates
Charlotte, North Carolina

Ralph Gerra
Lord, Day & Lord
New York, New York

William N. Kelly
Northwest Bancor Corporation
Minneapolis, Minnesota

Will S. Myers, Jr.
National Education Association
Washington, D.C.

James Harry Michael, Jr.
U.S. District Judge
Charlottesville, Virginia

H. Clyde Reeves
Chairman, Tax Policy Roundtable
Frankfort, Kentucky

Joel Stern
Stein, Stewart, Putnam,
  Macklis, Ltd.
New York, New York

Frederick D. Stocker
The Ohio State University
Columbus, Ohio

Deil S. Wright
University of North Carolina
Chapel Hill, North Carolina

About the Contributors

Richard J. Carlson is director of the Illinois Environmental Protection Agency (IEPA). Carlson was a special assistant to Governor Thompson from 1977 until he assumed the IEPA post in 1981. He has been director of research for both the National Municipal League and the Council of State Governments. Carlson has received the Ph.D. in political science from the University of Illinois and is a coauthor of *The Illinois Legislature.*

William J. Carroll is an economist with the American Telephone and Telegraph Company, doing demand analysis and forecasting for the interstate market. He is the author of several articles on public finance and regional economics topics. He has also been assistant professor of economics at Drew University and was on the staff of the Pennsylvania State University Land and Water Resources Center.

Charles W. Carry is the assistant chief engineer and assistant general manager of the Los Angeles Sanitation Districts. He holds the B.S. in civil engineering from Loyola University in Los Angeles and the M.S. in environmental engineering from the California Institute of Technology. From 1973 through 1976, he served as director of technology assessment for the National Commission of Water Quality. He has written numerous articles on wastewater treatment and solid waste management.

Stephen P. Coelen is associate professor in the School of Management at the University of Massachusetts, Amherst. He is also the director of the Massachusetts Development Research Institute. He has taught at The Pennsylvania State University and the University of Tennessee and was a senior economist at Abt Associates in Cambridge. He holds the Ph.D. in economics from Syracuse University and has published work on water resources, regional economics, demography, and econometrics.

Doris Van Dam is superintendent of the Grand Haven-Spring Lake, Michigan, Wastewater Treatment Plant. She is also a director of the Water Pollution Control Federation and a member of the executive board of the Michigan Water Pollution Control Association. Her published works include in-plant studies of the Grand Haven Plant as well as of Grand Rapids, Michigan, where she was assistant superintendent for twenty-eight years. She served on the management advisory group to the Environmental Protection Agency.

T. James Fries is executive staff advisor to the secretary of the Kentucky Natural Resources and Environmental Protection Cabinet. He has served as the chief of the Kentucky Division of Water's Planning and Standards Branch, as a river basin planner with the former Missouri River Basin Commission, and as water resources specialist on numerous contractual studies. He is also Kentucky's representative to the Water Management Subcommittee of the National Governor's Association. He holds the B.S. in natural resources administration and the M.S. degree in land and water conservation and management, both from Michigan State University.

Clair P. Guess, Jr., is a water management consultant who served for sixteen years as executive director of the South Carolina Water Resources Commission. As the commission's first director, he helped shape water resource policy for the state. Mr. Guess has served on the Interstate Conference on Water problems as chairman and as a member of the board of directors. He holds a B.S. from Clemson University in agricultural engineering.

Paul N. Guthrie, Jr., is director, Office of Intergovernmental Programs, Wisconsin Department of Natural Resources. He is a graduate of Swarthmore College and has studied at the Fels Institute of Local and State Government of the University of Pennsylvania. He has held a variety of posts in state and local government in several states and has consulted with numerous federal agencies. At present one of his principal assignments is managing the Wisconsin state municipal wastewater construction program, which incorporates both federally financed projects and projects developed under the state program.

Roger A. Kanerva is manager of environmental programs for the Illinois Environmental Protection Agency. Previously he was manager of the agency's Division of Water Pollution Control. He has also worked for the Water Resources Administration, Maryland Department of Natural Resources. He received the B.S. and M.S. in watershed management from the University of Arizona.

J. Leonard Ledbetter is director, Environmental Protection Division, for the Georgia Department of Natural Resources, and is a member of the Georgia Coastal Marshlands Protection Committee. He is a member of the Association of State and Interstate Water Pollution Control Administrators; Conference of State Sanitary Engineers; Water Pollution Control Federation; American Society of Civil Engineers; and Georgia Water and Pollution Control Association. He is a diplomate of the American Academy of Environmental Engineers. He received the B.S. and M.S. in engineering from Georgia Tech and has been associated with Georgia EPD since 1965.

James O. Mason received the M.D. from the University of Utah in 1958 and the Ph.D. in public health from Harvard University in 1967. He joined the U.S. Public Health Services as an epidemic intelligence service officer at the Center for Disease Control in Atlanta, in 1960. He concluded his service with the U.S. Public Health Service as deputy director for the Center and returned to Utah in 1970. Prior to his present position as executive director of the Utah Department of Health, he was commissioner of health services for the Church of Jesus Christ of Latter Day Saints and following that, associate professor and chairman of the Division of Community Medicine at the University of Utah College of Medicine.

Robert P. Miele heads the Technical Services Department of the Los Angeles County Sanitation Districts. He holds the B.S. and M.S. in civil sanitary engineering from The Pennsylvania State University. He has published articles on a wide variety of subjects in the wastewater treatment field including water reclamation, sludge processing and disposal, and advanced waste treatment.

George E. Peterson is director of the Public Finance Center of the Urban Institute. He is coauthor of *The Future of America's Capital Plant* (forthcoming). Peterson is a former Rhodes Scholar and did his graduate work in economics at Harvard University.

Mavis Mann Reeves is associate professor of government and politics, University of Maryland, College Park. She received the Ph.D. in political science from the University of North Carolina. She is author or coauthor of several books and articles on U.S. intergovernmental relations and state and local government including *Pragmatic Federalism: An Intergovernmental View of American Government.* She received the 1982 Donald C. Stone Award for "significant academic contribution to intergovernmental relations" from the American Society for Public Administration's Section on Intergovernmental Administration and Management.

Robbi J. Savage is executive director and secretary-treasurer of the Association of State and Interstate Water Pollution Control Administrators. As head of the Washington Office, she represents the state water officials on a number of major state and federal agency committees. She has written or has had management responsibility for several water quality documents, among them the *ASIWPCA Recommendations for an Improved National Water Quality Program (May, 1980)* and *States' Assessment of Progress under the Clean Water Act (August, 1981).* Beginning in the EPA's Water Planning Division, her environmental background includes work as a water quality specialist for the League of Women Voters Education Fund, an environmental analyst for the National Association of Manufacturers, and an independent consultant working with Linton, Mields, Reisler, and Cottoni.

Jacqueline A. Swigart is secretary of the Kentucky Natural Resources and Environmental Protection Cabinet. Prior to her appointment to the secretary's post in 1979, she was chairperson of the state's Environmental Quality Commission for seven years. She is the chairman of the Interstate Ohio River Basin Commission and vice chairman of the Interstate Conference on Water Problems (ICWP). She holds the B.S. in medical technology from the University of Minnesota and an honorary public service doctorate from the University of Louisville.

Joan K. Thomas is supervisor of water quality management for the Washington State Department of Ecology. She is a past president of the League of Women Voters of Washington and the Washington Environmental Council. She has chaired successful campaigns for three statewide water pollution control bond issues and has been involved in citizen support for clean water since 1958.

Richard Torkelson is assistant commissioner for administration in the New York State Department of Environmental Conservation. He is also an adjunct professor of public administration in the Graduate School of Government at Russell Sage College, Albany. He received the B.A. in scholastic philosophy from Saint Hyacinth College and the M.P.A. in administrative management from the University of Albany, where he is currently pursuing a D.P.A. His published works include coauthored articles on the management and policy implications of the hazardous waste problem.

John W.L. White is chairman and chief executive officer of Consumers Water Company in Portland, Maine. He attended Dartmouth College and received the B.S. in business and engineering administration from Massachusetts Institute of Technology in 1944. He is a registered professional engineer in four states and presents rate-of-return testimony before public utilities commissions. He is a past president of the National Association of Water Companies.

About the Editors

H. Clyde Reeves is chairman of the Tax Policy Roundtable of the Lincoln Institute of Land Policy and a member of the Kentucky Council of Economic Advisors. He was a long-time commissioner of revenue of Kentucky and has taught public-finance-related subjects at the University of Kentucky, the University of Louisville, and the University of Alabama in Huntsville, where he was the executive vice president. He is a past president of the National Association of State Tax Administrators and the National Tax Association—Tax Institute of America and was director of research of the Council of State Governments (1973–1975). He has been a consultant to the United Nations, Scientific Educational and Cultural Organization, and various federal government agencies, states, and industries. Reeves edited the *Role of the State in Property Taxation* and has published articles in the *National Tax Journal, Property Tax Journal,* and *Public Administration Review.*

Scott Ellsworth, a Washington-based writer and historian, has served as an editorial consultant to the National Academy of Public Administration, the Lincoln Institute of Land Policy, and other public policy organizations. He received the Ph.D. in history from Duke University.